A Baker's Dozen

A Baker's Dozen
Real Analog Solutions for Digital Designers

by **Bonnie Baker**

AMSTERDAM • BOSTON • HEIDELBERG • LONDON
NEW YORK • OXFORD • PARIS • SAN DIEGO
SAN FRANCISCO • SINGAPORE • SYDNEY • TOKYO

Newnes is an imprint of Elsevier

Newnes is an imprint of Elsevier
30 Corporate Drive, Suite 400, Burlington, MA 01803, USA
Linacre House, Jordan Hill, Oxford OX2 8DP, UK

Copyright © 2005, Elsevier Inc. All rights reserved.

No part of this publication may be reproduced, stored in a retrieval system, or transmitted in any form or by any means, electronic, mechanical, photocopying, recording, or otherwise, without the prior written permission of the publisher.

Permissions may be sought directly from Elsevier's Science & Technology Rights Department in Oxford, UK: phone: (+44) 1865 843830, fax: (+44) 1865 853333, e-mail: permissions@elsevier.com.uk. You may also complete your request on-line via the Elsevier homepage (http://elsevier.com), by selecting "Customer Support" and then "Obtaining Permissions."

 Recognizing the importance of preserving what has been written, Elsevier prints its books on acid-free paper whenever possible.

Library of Congress Cataloging-in-Publication Data

Baker, Bonnie.
  A Baker's dozen : real-world analog solutions for digital designers / by Bonnie Baker.
    p. cm.
  ISBN 0-7506-7819-4
    1. Digital integrated circuits--Design and construction. 2. Logic design. I. Title.

TK7874.B343 2005
621.3815--dc22   2005040558

British Library Cataloguing-in-Publication Data
A catalogue record for this book is available from the British Library.

For information on all Newnes publications
visit our Web site at www.books.elsevier.com

Transferred to Digital Printing in 2009.

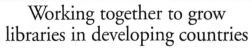

Contents

Preface .... *ix*
Acknowledgments .... *xi*
About the Author .... *xiii*

Chapter 1: Bridging the Gap Between Analog and Digital 1
Try to Measure Temperature Digitally 6
Road Blocks Abound 8
The Ultimate Key to Analog Success 14
How Analog and Digital Design Differ 15
Time and Its Inversion 20
Organizing Your Toolbox 21
Set Your Foundation and Move On, Out of the Box 22
Chapter 1 References 23

Chapter 2: The Basics Behind Analog-to-Digital Converters 25
The Key Specifications of Your ADC 28
Successive Approximation Register (SAR) Converters 40
Sigma-Delta (Σ–Δ) Converters 46
Conclusion 59
Chapter 2 References 60

Chapter 3: The Right ADC for the Right Application 63
Classes of Input Signals 65
Using an RTD for Temperature Sensing: SAR Converter or Sigma-Delta Solution? 72
RTD Signal Conditioning Path Using the Sigma-Delta ADC 76
Measuring Pressure: SAR Converter or Sigma-Delta Solution? 77
The Pressure Sensor Signal Conditioning Path Using a SAR ADC 79
Pressure Sensor Signal Conditioning Path Using a Sigma-Delta ADC 80
Photodiode Applications 81
Photosensing Signal Conditioning Path Using a SAR ADC 81
Photosensing Signal Conditioning Path Using a Sigma-Delta ADC 82
Motor Control Solutions 83

Conclusion 88
Chapter 3 References 89

Chapter 4: Do I Filter Now, Later or Never?.. 91
Key Low-Pass Analog Filter Design Parameters 95
Anti-Aliasing Filter Theory 103
Analog Filter Realization 105
How to Pick Your Operational Amplifier 108
Anti-Aliasing Filters for Near DC Analog Signals 109
Multiplexed Systems 112
Continuous Analog Signals 114
Matching the Anti-Aliasing Filter to the System 115
Chapter 4 References 116

Chapter 5: Finding the Perfect Op Amp for Your Perfect Circuit 117
Choose the Technology Wisely 121
Fundamental Operational Amplifier Circuits 122
Using these Fundamentals 129
Amplifier Design Pitfalls 131
Chapter 5 References 133

Chapter 6: Putting the Amp Into a Linear System 135
The Basics of Amplifier DC Operation 137
Every Amplifier is Waiting to Oscillate, and Every Oscillator is
   Waiting to Amplify 151
Determining System Stability 157
Time Domain Performance 161
Go Forth 163
Chapter 6 References 164

Chapter 7: SPICE of Life 165
The Old Pencil and Paper Design Process 172
Is Your Simulation Fundamentally Valid? 175
Macromodels: What Can They Do? 179
Concluding Remarks 183
Chapter 7 References 184

Chapter 8: Working the Analog Problem From the Digital Domain 185
Pulse Width Modulators (PWM) Used as a Digital-to-Analog Converter 188
Using the Comparator for Analog Conversions 194

Window Comparator 196
Combining the Comparator with a Timer 197
Using the Timer and Comparator to Build a Sigma-Delta A/D Converter 199
Conclusion 207
Chapter 8 References 208

Chapter 9: Systems Where Analog and Digital Work Together 209

Selecting the Right Battery Chemistry for Your Application 212
Taking the Battery Voltage to a Useful System Voltage... 213
Defining Power Supply Efficiency 214
Comparing The Three Power Devices 219
What is the Best Solution for Battery-Operated Systems? 221
Designing Low-Power Microcontroller Systems is a State of Mind 222
Conclusion 228
Chapter 9 References 229

Chapter 10: Noise – The Three Categories: Device, Conducted and Emitted .. 231

Definitions of Noise Specifications and Terms 234
Device Noise 238
Conducted Noise 254
Chapter 10 References 260

Chapter 11: Layout/Grounding (Precision, High Speed and Digital) 261

The Similarities of Analog and Digital Layout Practices 263
Where the Domains Differ – Ground Planes Can Be a Problem 266
Where the Board and Component Parasitics Can Do the Most Damage........... 267
Layout Techniques That Improve ADC Accuracy and Resolution 274
The Art of Laying Out Two-Layer Boards 277
Current Return Paths With or Without a Ground Plane 281
Layout Tricks for a 12-Bit Sensing System 282
General Layout Guidelines – Device Placement 284
General Layout Guidelines – Ground and Power Supply Strategy 284
Signal Traces.. 287
Did I Say Bypass and Use an Anti-Aliasing Filter?.. 287
Bypass Capacitors... 287
Anti-Aliasing Filters 288
PCB Design Checklist 288
Chapter 11 References 290

Chapter 12: The Trouble With Troubleshooting Your Mixed-Signal Designs Without the Right Tools 291

The Basic Tools for Your Troubleshooting Arsenal 293
You ask, "Does my Circuit A/D Converter Work?" 295
Power Supply Noise 298
Improper Use of Amplifiers 301
Don't Miss the Details 303
Conclusion 305
Chapter 12 References 306

Chapter 13: Combining Digital and Analog in the Same Engineer, and on the Same Board 307

The Signal Chain to the Real World 309
Tools of the Trade 310
Throwing the Digital In With the Analog 314
Conclusion 318

Appendix A: Analog-to-Digital Converter Specification Definitions and Formulas 319

Appendix B: Reading FFTs 329

Reading the FFT Plot 331

Appendix C: Op Amp Specification Definitions and Formulas 337

Index 343

Preface

I went to an analog university where the core courses were, of course, all analog. Then I started my career at a high-quality, premier, analog house; Burr-Brown. Mind you, my objective was not to work at an analog house, my objective was to have a job. Nonetheless, there I rubbed shoulders with the best analog engineers in industry. After thirteen years, I decided to expand my horizons and work for a digital company. For me, thirteen was a lucky number because this is when my real education began.

What did I learn? I learned that you design your circuit so that it works in the application, not so that you have the most elegant solution in industry. I also learned that you can use digital circuits as well as analog circuits to get the job done. Moreover, I learned that sometimes ignorance is bliss. Many of the digital engineers that I have worked with don't know that some tasks are impossible. For instance, at Burr-Brown we trimmed our precision analog circuits with the high technology of Nicrome. This trim process is very specific to analog silicon circuits and is accurate. I told the engineers that they could not have precision products without a Nicrome process. Boy was I wrong. Microchip trims in analog circuit precision with their digital Flash process.

I have always been a "died in the wool" analog engineer, but I am starting to change. I haven't made a total transition to the "dark (digital) side," but digital is looking more attractive all the time. This attraction is enhanced by the fact that I am very familiar with analog and have a diverse set of analog, and now digital, tools to solve my circuit problems. This book is for you so that you can also have the same set of tools and can become more competitive in your design endeavors.

Digital circuitry and software is encroaching into the analog hardware domain. Analog will never disappear at the sensor conditioning circuit, power supply, or layout strategies. I know the digital engineer will continue to be challenged by analog issues, even if they deny that they exist.

Now let's add to the complexity of the digital engineer's challenges. The advances in microcontroller and microprocessor chip designs are growing in every direction. Increased speed and memory is just one example of the direction that these devices are taking. However, the most interesting change is the addition of peripherals, including analog and interface circuitry. Not only is the engineer required to know the details of the implementation of these peripherals,

Preface

but also know the basics of layout strategies. Today, the digital engineer needs an added dimension of knowledge in order to solve problems beyond the firmware design challenges.

Going forward, the digital engineer needs some basic tools in their toolbox. I wrote this book for practicing digital engineers, students, educators and hands-on managers who are looking for the analog foundation that they need to handle their daily engineering problems. It will serve as a valuable reference for the nuts-and-bolts of system analog design in a digital word. The target audience for this book is the embedded design engineer that has the good fortune to wander into the analog domain.

Acknowledgments

I'd like to thank all of the engineers who gave their time to review the material in the volume. A primary reviewer, Kumen Blake (Microchip Technology engineer) was meticulous and always provided excellent, relevant feedback. Paul McGoldrick (AnalogZone, editor-in-chief) gave significant time to ensure that sections of this book were accurate and concisely written. Numerous engineers at Microchip Technology, Texas Instruments and Burr-Brown also reviewed the material for technical accuracy.

Thanks also to Newnes acquisition editor Harry Helms and Kelly Johnson of Borrego Publishing. Harry pestered me for over a year to just sit down and write. I then said to him it would take two years to finish this book, and he said it would take one year. It actually took ten months from start to finish only because of Harry's enthusiastic encouragement at the beginning. Kelly did an outstanding job of editing my final-author's copy.

And especially, thanks to my support system in Tucson, Arizona. They were my cheerleaders in this solitary endeavor. And together, we finished it!

About the Author

Bonnie Baker writes the monthly "Baker's Best" for *EDN* magazine. She has been involved with analog and digital designs and systems for nearly 20 years. Bonnie started as a manufacturing product engineer supporting analog products at Burr-Brown. From there, Bonnie moved up to IC design, analog division strategic marketer, and then corporate applications engineering manager. In 1998, she joined Microchip Technology and has served as their analog division analog/mixed-signal applications engineering manager and staff architect engineer for one of their PICmicro divisions. This has expanded her background to not only include analog applications, but microcontroller solutions as well.

Bonnie holds a Masters of Science in Electrical Engineering from the University of Arizona (Tucson, AZ) and a bachelor's degree in music education from Northern Arizona University (Flagstaff, AZ). In addition to her fascination with analog design, Bonnie has a drive to share her knowledge and experience and has written over 200 articles, design notes, and application notes and she is a frequent presenter at technical conferences and shows.

CHAPTER 1

Bridging the Gap Between Analog and Digital

CHAPTER 1

Bridging the Gap Between Analog and Digital

A few years ago, I was approached by a new graduate, engineering applicant at the Embedded Systems Conference (ESC), 2001 in San Francisco. When he found out that I was a manager, he explained that he was looking for a job. He said he knew of Microchip Technology, Inc. and wanted to work for them if he could. He immediately produced his resume. I gave him a few more details about my role at Microchip. At the time, I managed the mixed signal/linear applications group. My department's roles were product definition, technical writing, customer training, and traveling all over the world visiting customers. At the conclusion of this "sales" pitch, he proudly told me that it sounded like a great job. I reemphasized that I was in the Analog arm at Microchip. He obviously thought that he did his homework because he told me that analog is dying and digital will eventually take over. Anyone who knew anything about Microchip would agree! Wow, I had a live one.

I was there, doing my obligatory Microchip booth duty for the afternoon. There was a lot of action on the floor, and the room was full of exhibits. The lights were on, the sound of conversations were projecting across the room. The heating and cooling system was doing a splendid job of keeping us comfortable. Exhibitors in the booths were (believe it or not) demonstrating the operation of sensors, power devices, passive devices, RF products, and so forth. There must have been several hundred booths, all of which were trying to promote their engineering merchandise.

Figure 1.1: The Embedded Systems Conference exhibit hall in 2001 had hundreds of booths, many of which were already showing signs of interest in analog systems. This was done even though the emphasis of the conference was digital.

Some of the vendor exhibits had analog signal conditioning demonstrations. As a matter of fact, right in front of us, Microchip had a temperature sensor connected to a computer through the parallel port. The temperature sensor board would self-heat, and the sensor would measure this change and show the results on the PC screen. Once the temperature reached a threshold of 85°C, the heating element was turned off. You could then watch the temperature go down on the PC until it reached 40°C, at which point the heating element would be turned-on again.

At a second counter, we also had a computer running the new FilterLab® analog filter design program. With this tool, you can specify an analog filter in terms of the number of poles, cut-off frequency and approximation type (Butterworth, Bessel and Chebyshev). Once you type in your information, the software spits out a filter circuit diagram, such as the filter circuit shown in Figure 1.2. You can theoretically build the circuit and take it to the lab for testing and verification. There was a customer at the counter, playing around with the filter software.

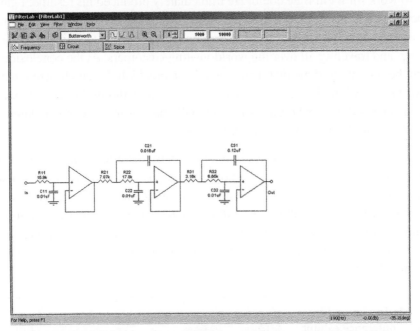

Figure 1.2: One of the views of the FilterLab program from Microchip provided analog filter circuit diagrams. This particular circuit is a 5th order, low-pass Butterworth filter with a cut-off frequency of 1 kHz. The FilterLab program from Microchip is just one example of a filter program from a semiconductor supplier. Texas Instruments, Linear Technology, and Analog Devices have similar programs available on the World Wide Web.

At exhibit counter number three, there was a CANbus demonstration with temperature sensing, pressure sensing and DC motor nodes. CANbus networks have been around for over 15 years. Initially, this bus was used in automotive applications requiring predictable, error-free

Bridging the Gap Between Analog and Digital

communications. Recent falling prices of controller area network (CAN) system technologies have made it a commodity item. The CANbus network has expanded past automotive applications. It is now migrating into systems like industrial networks, medical equipment, railway signaling and controlling building services (to name a few). These applications are using the CANbus network, not only because of the lower cost, but because the communication with this network is robust, at a bit rate of up to 1 Mbits/sec.

A CANbus network features a multimaster system that broadcasts transmissions to all of the nodes in the system. In this type of network, each node filters out unwanted messages. An advantage from this topology is that nodes can easily be added or removed with minimal software impact. The CAN network requires intelligence on each node, but the level of intelligence can be tailored to the task at that node. As a result, these individual controllers are usually simpler, with lower pin counts. The CAN network also has higher reliability by using distributed intelligence and fewer wires.

You might say, "What does this have to do with analog circuits?" And the answer is *everything*. The communication channel is important only because you are shipping digitized analog information from one node to another. With this ESC exhibit, three CANbus nodes communicated through the bus to each other. One node measured temperature. The temperature value was used to calibrate the pressure sensor on the second node. You could apply pressure to the pressure-sensing node by manually squeezing a balloon. (This type of demonstration was put together to get the observer more involved.) The sensor circuitry digitized the level pressure applied and sent that data through the CANbus network to a DC motor. The DC motor was configured so that increased pressure would increase the revolution per minute (RPM) of the motor. Figure 1.3 shows a basic block diagram containing the pressure-sensing node.

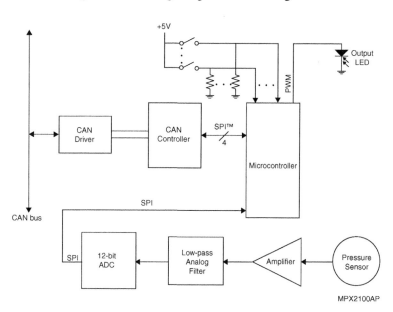

Figure 1.3: The CANbus system at the 2001 Embedded Systems Conference has three different analog function nodes. The node illustrated in this figure measured the pressure applied to a balloon and sent the data across the CANbus network to a DC motor (not illustrated here).

Chapter 1

Then to finish out the Microchip displays in the booth, there were three counters that had microcontroller demos.

I asked the engineering applicant, giving him a chance to redeem himself, "Out of curiosity, do you see anything analog-ish like in this room?" He looked around the convention room thoughtfully. I was amused when he sympathetically looked at me and answered, "No, not really." I think that he thought I was a bit old-fashioned, behind the times. No regrets from him. He was confident that he gave me an insightful, informed answer.

You guessed it. His resume went into the circular file.

Try to Measure Temperature Digitally

No, this is not a book about interview techniques. This book is neither about how to win points and climb the corporate ladder. This is a book about the analog design opportunities that surround us every day, all day long, and how we can solve them in a single-supply environment. Reflecting on the applicant's answer, I think that he was partially right. Digital solutions are encroaching into the analog hardware in a majority of applications.

So let's try to measure temperature digitally. The simple, low resolution analog-to-digital (A/D) converter can easily be replaced with a resistor/capacitor (R/C) pair connected to a microcontroller I/O pin. The R/C pair would supply a signal that operates with a single-pole, rise-time function. The controller counts milliseconds, with its oscillator/timer combination measures the input signal. Why would you want to do this? Maybe you are measuring temperature with a sensor that changes its resistance value with changes in temperature.

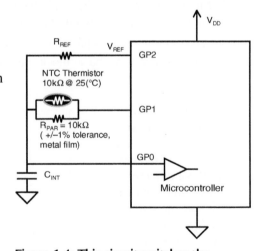

The temperature sensing circuit in Figure 1.4 is implemented by setting GP1 and GP2 of the microcontroller as inputs. Additionally, GP0 is set low to discharge the capacitor, C_{INT}. As the voltage on C_{INT} discharges, the configuration of GP0 is changed to an input and GP1 is set to a high output. An internal timer counts the amount of time (t_1 in Figure 1.5) before the voltage at GP0 reaches the threshold (V_{TH}), which changes the recognized input from 0 to 1. In this case, R_{NTC} (a negative temperature coefficient thermistor) is placed in parallel with R_{PAR} or $R_{NTC} \| R_{PAR}$. This parallel combination interacts with C_{INT}. After this happens, GP1 and GP2 are again set as inputs and

Figure 1.4: This circuit switches the voltage reference on and off at GP1 and GP2. In this manner, the time constant of the NTC thermistor in parallel with a standard resistor (RNTC || RPAR) and integrating capacitor (C_{INT}) is compared to the time constant of the reference resistor (R_{REF}) and integrating capacitor.

Figure 1.5: The R/C time response of the circuit shown in Figure 1.4 allows for the microcontroller counter to be used to determine the relative resistance of the negative temperature coefficient (NTC) thermistor element.

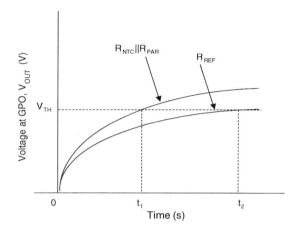

GP0 as an output low. Once the integrating capacitor C_{INT} has time to discharge, GP2 is set to a high output and GP0 as an input. A timer counts the amount of time before GP0 changes to 1 again, but this provides the timed amount of t_2, per Figure 1.5. In this case, R_{REF} is the component interacting with C_{INT}.

The integration time of this circuit can be calculated using:

$V_{OUT} = V_{REF} (1 - e^{-t/RC})$ or

$t = RC \ln (1 - V_{TH} / V_{REF})$

where V_{OUT} is the voltage at the I/O pin, GP0,

V_{REF} is the output, logic-high voltage of the I/O pin, GP1 or GP2;

V_{TH} is the input voltage to GP0 that causes a logic 1 to trigger in the microcontroller.

If the ratio of $V_{TH}:V_{REF}$ is kept constant, the unknown resistance of the $R_{NTC} \| R_{PAR}$ can be determined with:

$$R_{NTC} \| R_{PAR} = R_{REF} (t_2/t_1)$$

Notice that in this configuration, the resistance calculation of the parallel combination of $R_{NTC} \| R_{PAR}$ is independent of C_{INT}, but the absolute accuracy of the measurement is dependent on the accuracy of your resistors.

Oops, did I say you can use a linear resistor and a charging device like a capacitor to replace an A/D converter in a temperature measurement system? I guess my applicant at the ESC show was also wrong. Analog will never disappear and the digital engineer will continue to be challenged to delve into these types of issues. The analog solution is many times more efficient and usually more accurate. For instance, the previous R/C example is only as accurate as the number of bits in the timer, the speed of the oscillator, and how accurately you know the value of your resistors.

Chapter 1

Road Blocks Abound

I have worked with a wide spectrum of analog and digital designers. Each one of them has their own quirks and reasons why they can't do everything, but here are some statements that I have received from my digital clientele concerning their analog challenges.

Not My Job!

This statement came about with surprising frankness. "People in my department are avoiding analog circuitry in their design as much as possible, no matter how important it is. Many of them have had experiences where analog circuit performance was hard to predict. Therefore, almost every engineer will find an existing analog circuit and use that as a point of reference. If they have the misfortune of being asked to design part or all of the analog circuit from scratch, they will try to use facts that they remember from their school days. And in their school days they studied mostly digital."

Good luck. It seems from this statement that the died-in-the-wool digital designer has no interest in how to get from A to B, but more interest in what the cookbook suggests.

It turns out that the designers who operate in this mode are like a carpenter with a hammer looking for a nail. The designer has a circuit solution and tries to make it fit their application. A good example of applying the cookbook solution to a place where it won't fit is to try to use a standard 12-bit successive approximation register (SAR) in a power sensing application. This type of application actually requires a sigma-delta converter. As you will find later in this book (Chapter 3), the sigma-delta ($\Sigma-\Delta$) converter can reach a resolution level in the sub-nano volt region. This is an advantage because you not only eliminate the input, analog-gain stage, but you reduce the noise in the bandpass region of your signal. Figure 1.6 shows this power meter solution.

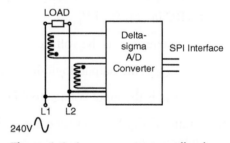

Figure 1.6: A power meter application requires <12-bit resolution in the system. This may imply that a simple 12-bit SAR converter can do the job, but the required LSB size is much smaller than the SAR converter can provide. Consequently, a sigma-delta converter is often used.

In this circuit, the current through the power line is sensed using an inductor on the low-side of the load. As a result, the voltage drop across this sensing element must be low.

Show Me the Beef

One day, a digital engineer said to me, "Thank god, I have finally found the key to working with analog and now I can go back about my digital business. Thank you for that one, insightful tip."

The tip I gave him was not that earthshaking. As a matter of fact, it provided the two primary operational amplifier specifications that an engineer uses when designing an analog low-pass filter. See Gain Bandwidth Product and Slew Rate (Chapter 4).

The gain bandwidth product (GBWP) is a multiplication factor that helps you predict the closed-loop bandwidth of an operational amplifier. You can easily find this parameter by looking at the specification table of the amplifier. You can quickly find this specification out of the 30 or 40 items in the table by looking at the "units" column. That column is usually on the right side of the table. When you are trying to find the gain bandwidth product specification, look for frequency units in Hz, kHz or MHz. Once you find these abbreviated units, verify that you have found the right item by looking to the left for the specification definition. Now, double-check and ensure you understand the test conditions for this specification by reading the conditions column and the general conditions that are summarized at the top of the table. All of these areas on a typical data sheet are pointed out in Figure 1.7.

AC ELECTRICAL SPECIFICATIONS

Electrical Characteristics: Unless otherwise indicated, T_A = +25°C, V_{DD} = +1.8 to 5.5V, V_{SS} = GND, V_{CM} = V_{DD}/2, V_{OUT} = V_{DD}/2, R_L = 10 kΩ to V_{DD}/2, and C_L = 60 pF.

Parameters	Sym	Min	Typ	Max	Units	Conditions
AC Response						
Gain Bandwidth Product	GBWP	—	1.0	—	MHz	
Phase Margin	PM	—	90	—	°	G = +1
Slew Rate	SR	—	0.6	—	V/µs	
Noise						
Input Noise Voltage	E_{ni}	—	6.1	—	µVp-p	f = 0.1 Hz to 10 Hz
Input Noise Voltage Density	e_{ni}	—	28	—	nV/√Hz	f = 1 kHz
Input Noise Current Density	i_{ni}	—	0.6	—	fA/√Hz	f = 1 kHz

Figure 1.7: A typical electrical specification table for an operational amplifier has seven columns. When searching for a particular specification, the units-column is the easiest one to start with.

A second place where this specification can be found is in the characteristic performance graphs later on in the data sheet (see Figure 1.8). *Open-loop gain versus frequency* is the usual label for this curve. Sometimes the open-loop phase is included in this graph. You will find that the 0dB crossing of the gain curve will usually match the gain bandwidth product in the specification table.

Figure 1.8: These typical performance curves show many of the parameters in the specification table of the data sheet of an amplifier. This graph illustrates the typical open-loop gain, phase vs. frequency response. The arrow in this figure points to the gain bandwidth product for this unity-gain-stable amplifier.

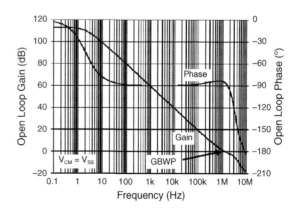

Chapter 1

The gain bandwidth product (GBWP) will tell you the highest small-signal frequency (~ ±100 mV) that you can send through your amplifier circuit without distortion. It also tells you the frequency where a pole is introduced to your closed-loop amplifier circuit. This is particularly critical to know when you design low-pass filters. In this type of circuit, you deliberately put poles in the transfer function by putting resistors and capacitors around the amplifier, as shown in Figure 1.2. If the amplifier adds a pole, your circuit could oscillate. Consequently, the closed-loop bandwidth of the amplifier must be at least 100 times higher than the cutoff frequency ($f_{CUT\text{-}OFF}$) of the filter. Another way of stating this is that the gain bandwidth product of your amplifier should be equal to or greater than $100 \times f_{CUT\text{-}OFF}$ (this assumes the filter has a gain of +1 V/V). If you don't take these precautions, you might erroneously be inclined to investigate your filter equations only to find out that the amplifier is not well-suited for your design.

You might ask, "How important is this specification in other amplifier application circuits?" Generally, you will need an amplifier with good bandwidth performance for your signal, but probably won't see instability because of your amplifier selection. Or in another application, you may be more concerned about the quiescent current of the amplifier or power supply capability instead of the bandwidth because you are designing a battery-powered circuit that operates at DC.

The second specification that I mentioned previously is slew rate. The slew rate of an amplifier is determined by putting a square wave signal at the input of the amplifier and looking to see how fast the signal changes on the output. The units of this specification are generally V/sec, V/msec, or V/μsec. You can find this specification in the table in the same way we found the gain bandwidth product. There is also a characteristic curve in most amplifier data sheets that gives a good look at how a typical part will perform. You'll find that the label of this graph is usually "Large signal, noninverting pulse response" (Figure 1.9).

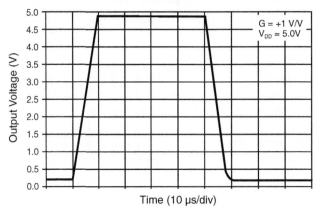

Figure 1.9: This graph illustrates the typical time domain response of the output voltage vs. time of an amplifier.

With the filter circuit, this specification will tell you the maximum frequency of the large signals going through your circuit. If you don't pay attention to this specification, you may find that the amplifier distorts your larger, higher frequency signals. A good rule of thumb for this design parameter is: slew rate $\geq (2\pi V_{OUT\,P\text{-}P} \times f_{CUT\text{-}OFF})$ where $V_{OUT\,P\text{-}P}$ is the expected peak-to-peak output voltage swing below $f_{CUT\text{-}OFF}$ of your filter.

Once again, you may ask, is this specification critical in all applications? The answer is again, no. You will find that the operational amplifier applications are very diverse. As a result, op amp manufacturers average 30 to 40 specifications in the tables and 15 to 25 characteristic curves. This is done to cater to as many users as they can. It is useful to note that the gain bandwidth product and slew rate were primary specifications for this one type of circuit. Meeting these specification requirements is critical if you are designing a low-pass filter, but this is not the case with other operational amplifier applications.

Don't Bother Me With the Small Stuff—Just Give Me the Data

One of the more common statements as said to me by the ambitious, digital engineer is, "Just give me the data. I will fix it in my processor. I know we can design a digital filter with the classical FIR or IIR filters, or better yet implement an FFT response. I can also calibrate the signal if need be. I'm confident that I will be able to get rid of those undesirable, messy analog signals."

This comment always brings a smile to my face. See the case in point with the circuit in Figure 1.10.

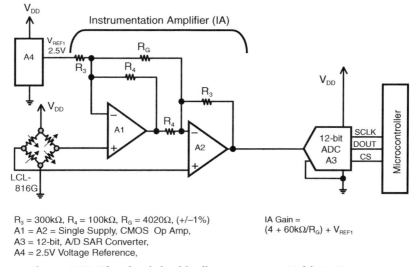

$R_3 = 300k\Omega$, $R_4 = 100k\Omega$, $R_G = 4020\Omega$, (+/–1%)
A1 = A2 = Single Supply, CMOS Op Amp,
A3 = 12-bit, A/D SAR Converter,
A4 = 2.5V Voltage Reference,

IA Gain =
$(4 + 60k\Omega/R_G) + V_{REF1}$

Figure 1.10: The circuit in this diagram uses a 12-bit A/D converter in combination with an instrumentation amplifier to convert the low-signal, output of a Wheatstone bridge sensor to usable digital codes.

The analog portion of this circuit has a load cell, a dual-operational amplifier configured as an instrumentation amplifier, a SAR A/D converter, a microcontroller and voltage references for the IA and A/D converter. The sensor is a 1.2 kΩ, 2 mV/V load cell with a full-scale load range of ±32 ounces. In this 5 V system, the electrical full-scale output range of the load cell is ±10 mV. The instrumentation amplifier, consisting of two operational amplifiers (A1 and A2) and five resistors, is configured with a gain of 153 V/V. This gain matches the full-scale output swing of the instrumentation amplifier block to the full-scale input range of the A/D converter. The SAR A/D converter has an internal input sampling mechanism. With this function, a single sample is taken for each conversion. The microcontroller acquires the data from the SAR A/D converter. The controller can also execute calibration and translate the data into a usable format for tasks such as displays or actuator feedback signals.

The transfer function, from sensor to the output of the A/D converter is:

$$D_{OUT} = ((LC_P - LC_N)(Gain) + V_{REF1})(2^{12}/V_{REF2})$$

with $LC_P = V_{DD} (R_2 /(R_1 + R_2))$

with $LC_N = V_{DD} (R_1 /(R_1 + R_2))$

with $GAIN = (1 + R_3 /R_4 + 2R_3 /R_G)$

where LC_P and LC_N are the positive and negative sensor outputs,

GAIN is the gain of the instrumentation amplifier circuit. The instrumentation amplifier is configured using A1 and A2. The gain is adjusted with R_G,

V_{REF1} is a 2.5 V reference which level shifts the instrumentation amplifier output

V_{REF2} is the 4.096 V reference, which determines the A/D converter input range and LSB size;

V_{DD} is the power supply voltage and sensor excitation voltage;

D_{OUT} is a decimal representation of the 12-bit digital output code of the A/D converter (rounded to the nearest integer).

If the design of this system is poorly implemented, it could be an excellent candidate for noise problems. The symptom of a poor implementation is an intolerable level of uncertainty with the digital output results from the A/D converter. It is easy to assume that this type of symptom indicates that the last device in the signal chain generates the noise problem. But, in fact, the root cause of poor conversion results could originate with other active devices or passive components in the signal chain, the PCB layout or even extraneous sources.

In this circuit, noise can be reduced within the analog channel hardware. But, with the first prototype of this circuit, these low noise precautions were not used. Therefore, the data output of the A/D converter illustrated in Figure 1.11 indicates that this was a noisy system. It is fine

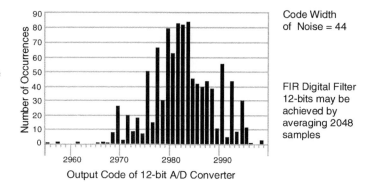

Figure 1.11: A poor implementation of the 12-bit data acquisition system shown in Figure 1.10 could easily have an output range of 44 different codes with a 2,048 sample size.

Code Width of Noise = 44

FIR Digital Filter 12-bits may be achieved by averaging 2048 samples

to design a proto with this level of noise. In addition, it is truly divine to understand the noise and remove it in hardware wherever possible.

But, let's assume that you take the digital route to perform the filtering. On a perfect day, you will need to collect at least 2,048, 12-bit data points and calculate the average. I said on *a perfect day* because when I look at this data there seems to be more going on than just white noise. There are small occurrences in the lower 20 codes of the data, and the major portion of the data does not form a "normal distribution" type of curve. It seems to have troughs and there is nothing normal about this data at all.

A common, bad scenario is that the problem is never solved through the lifetime of your application circuit. These unknown noise problems are fixed with digital tricks. That overly confident statement ignores the trade-offs inherent in taking the all-digital route. One of the major consequences is time. A digital filter needs to collect several hundreds of samples in order to compete with the analog solution. On top of that, the already digitized signal has been contaminated by aliased (this is explained in Chapter 4), high-frequency signals, which you will never be able to tell your original signal from the contaminants. These tricks may or may not work over time.

On the other hand, the analog solution is simple and final. The data loses its erratic behavior and you can get the same converted number every time! What do you do?

1. Put bypass capacitors across the power supply pin to ground with every active device. (See Chapter 10.)

2. Use a ground plane. This will usually require at least a two-layer board. (See Chapter 11.)

3. Reduce the resistor values in the instrumentation amplifier. When you reduce these resistors (without changing the throughput gain), the noise in the signal chain will also reduce. (See Chapter 10.)

4. Use low noise amplifiers. (See Chapter 12.)

Chapter 1

5. Insert a low-pass filter before the A/D converter. This filter will remove higher frequency noise as well as eliminate aliasing problems. (See Chapter 4.)

6. Choke the power supply switching noise to the analog portion of the board with an inductor. (See Chapter 12.)

These are all simple solutions to a seemingly impossible, noisy circuit problem. Figure 1.12 shows the results of these actions.

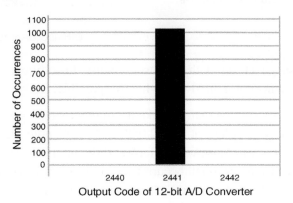

Figure 1.12: When noise reduction techniques are used in the implementation of the circuit in Figure 1.10, it is possible to get a 12-bit system.

Calibration can be another sticky point when you go to the digital environment. Once you loose your dynamic range in the analog domain, it is impossible to recover it digitally. For instance, if you use amplifiers in this circuit that do not give you good rail-to-rail performance, the outer limits of the signal is lost forever. Another situation, not related to Figure 1.10, could be if your signal is logarithmic instead of linear. If this is the case, digital manipulation will not take you very far. This type of data can only be fixed in the analog domain.

The Ultimate Key to Analog Success

News alert! The ultimate analog key does not exist. And I don't mind turning this around to tell analog engineers that digital engineering is a little more than ones and zeros. The analog mountains that can be climbed are analogous to your digital challenges. Following are three examples.

For the first example, in the sprit of designing a robust design, the digital designer architects the software to identify unforeseen, catastrophic errors. The watchdog timer (WDT) can be used for this purpose. The function of a watchdog timer is easy enough. It counts down using the system clock from an initial value to zero. During implementation, if your firmware does not reset this timer soon enough, the watchdog timer resets or interrupts the system without human intervention when the counter reaches zero. Alternatively, the analog domain protection circuitry is used to minimize the effects of unforeseen errors or transients. In analog disciplines this be can implemented with over-range notifications or protection devices at sensitive nodes, such as zener diodes, metal oxide varistors (MOV), transzorbes, or Schottky diodes. With these types of additions to the hardware, "bad" signals are identified and eliminated before they become part of the signal path.

The second example would be to work on your digital design low-power strategies by effectively using clocking algorithms. Low power should be thought of as a "state of mind" (see Chapter 9). With a low-power mindset, you can throttle down your controller to near inactivity if you really want to save battery power. The hardware approach would be to reduce clock source rate or power supply voltage. An equally effective approach is to operate with a partial or complete controller/processor shutdown mode. Combining these techniques with execution time and a little intelligence, you can easily tackle your most challenging power conservation problems. In your analog design, you will choose the lower power devices and utilize device shutdown features. In this environment, the designer needs to research the market for the best solution, whether it is a similar lower power device, or an alternative silicon topology that runs more efficiently.

A third example would be where you savor and protect your programming tricks from your competition. You can do this by making the code unreadable in the finished product in the same way the analog engineer buries traces inside boards, blacks out device part numbers, or asks vendors to give him proprietary part numbers.

The list goes on. But the thing to remember and understand is that each of us, in our own disciplines, tries to take technology to its limit. So what do you do when your manager says in his one-sided conversational way, "You're an engineer (aren't you)? Good. Since we are understaffed, I need you to do the entire (hardware/software) design. What? You don't know anything about analog. Hmmm, maybe I need to find someone else? I knew you would rise to the occasion. Have your development schedule on my desk by the end of the day so I can set up a deadline schedule."

How Analog and Digital Design Differ

The basic difference between the analog mindset and digital mindset are embedded in the definitions of precision (calculated risk versus right every time), hardware versus software, and time (or the inverse of). The basic concepts behind analog and digital disciplines are easy to find. In terms of this book, I will describe analog design from a practical standpoint. You will find that the in-depth lists and details about product specifications will be a little thin, but there is a detailed discussion about key specifications as they relate to basic analog systems.

Precision

What is precise enough in an analog circuit? There are three ways to answer this question. A first aspect is, "as precise as it needs to be." You will find that some of your circuits will only require accuracy to one or two millivolts. Others will require accuracy to the submicrovolts.

This difference in system requirements will encourage you to settle for "close enough" in some systems, and "What else can I squeeze out of this circuit?" in other systems.

A second aspect of accuracy involves really understanding the components and devices you are working with. In terms of the components, you should know that a 1 kΩ resistor or a 20 pF capacitor is not equal to those absolute values all the time. For instance, temperature can have a dramatic effect of these components. Also, there are variations from device-to-device out of your bin in the lab. The combination of these two major issues can change the performance of your circuit dramatically if you don't take them into consideration.

In terms of devices, you will find product data sheets have maximum guaranteed values and typical values. The maximum guaranteed values are self-explanatory in that you should expect that your devices will not over-range the specifications as stated provided the devices are not overstressed with higher voltages or temperatures. The typical values are another manner. There are a variety of ways to determine what these typical values should be, and you will find that each manufacturer will have their own way to calculate these values along with their justification. Some manufacturers take the average of a large sample of devices prior to the initial product release. Other manufacturers define their typical values as being equal to one standard deviation plus the average. I have also heard of manufacturers using their SPICE simulation as a guide for these numbers. Sometimes the Spice simulation is justified because it is impossible to test a particular specification.

The third aspect of accuracy is noise. When you take this issue into consideration, you need to have some understanding of statistical calculations with large samples. I am going to cover this issue in more detail later in Chapter 10.

Hardware versus Software

This discussion seems to simplify the problem a bit, but I have a solution for those embarking on the ownership of analog. Think of it in terms of learning the fundamentals about your components, knowing the general behavior of basics building block devices, and running through a high level evaluation of your circuits first.

1. *Learn the fundamentals about your components.*

For instance, the fundamentals at the very bottom of the barrel include resistors, capacitors and inductors. You were probably exposed to the devices early in your career, but what do you really need to know as an analog design engineer?

Resistors are simple devices. There are several perspectives that you have to consider when you use this type of component in your design. The first, and most easy way of thinking about a resistor is that it influences voltage and current in your design. This is defined through the infamous Thevenin equation:

V / RI = 1

Where V is voltage

R is resistance in ohms

I is current in amperes

I always remember this formula from my elementary school geography lessons. **V**ermont is always over **R**hode **I**sland.

But this is the part of resistors in your circuit. For practical purposes, this is a DC equation, not AC. Moving past this formula, you need to be concerned about the parasitic characteristics. Namely, there is a parasitic capacitor in parallel with the resistive element and a parasitic inductor in series. These components are artifacts of the physical device. There is a diagram of the resistor with these parasitics in Figure 1.13.

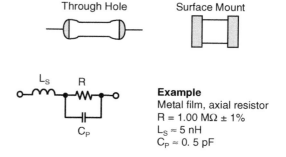

Figure 1.13: This illustrates a typical resistor model. The parasitic elements of a standard resistor are parallel capacitance (C_P) and series inductance (L_S).

Example
Metal film, axial resistor
R = 1.00 MΩ ± 1%
$L_S \approx$ 5 nH
$C_P \approx$ 0.5 pF

The fact is, I never worry about the parasitic capacitance until I started designing transimpedance, optical, photodiode-sensing circuits. An example of this type of circuit is shown in Figure 1.14. If blindly built (without concern for the parasitic capacitance), this photosensing circuit can mysteriously sing like a bird (oscillate) without too much effort. This oscillation is usually caused by an inappropriate choice of C_F, but it can also be caused by that phantom capacitor, C_P. These capacitors, in combination with the photodiode parasitic capacitance and the amplifier's

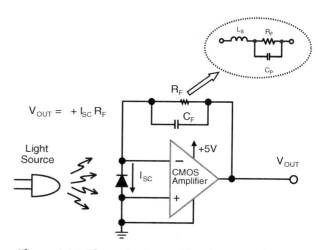

Figure 1.14: If you don't consider the parasitic capacitance of the feedback resistor, a transimpedance photosensing circuit can be unstable.

input capacitance interact to establish stability, or not (see Chapter 6). This is one example, but you can extrapolate this to other circuits if you are using small value discrete capacitors in parallel or series with discrete resistors.

The parasitic inductance of the resistor (also see Figure 1.15) can affect higher speed systems where lower value resistors are the norm. This inductance can affect the behavior of the current sensing resistor used in switched-mode power supplies.

Generally speaking, the impedance of higher value resistors is more affected by the parasitic capacitance, and low value resistors is affected by the parasitic inductance. Figure 1.15 illustrates this point.

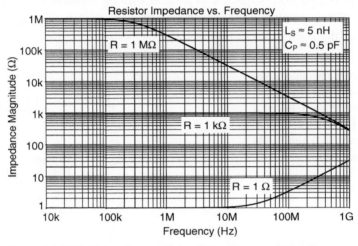

Figure 1.15: The impedance of a resistor changes from the defined DC resistance value to other values over frequency. The parasitic capacitance and impedance influence these changes.

Capacitors, on the other hand, should be considered in the frequency domain when you are designing. There is one formula for the capacitor that I used frequently in my design. This formula is:

$I = C * \delta V/\delta t$

Where C is capacitance in farads

δV is change in voltage in volts

δt is change in time is seconds

Capacitors are very useful for power supplies, stability, loading low dropout regulators and loading voltage references. But, in all cases, you use capacitors to modify frequencies, not DC signals.

Bridging the Gap Between Analog and Digital

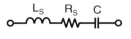

Example (signal-quality capacitor)
1206 SMT ceramic chip capacitor
X7R
$C = 1.0\ \mu F \pm 10\%$
$L_S \approx 3$ nH
$R_S \approx 0.01\ \Omega$

Figure 1.16: This illustrates a typical ceramic capacitor model. The parasitic elements of a standard capacitor is parallel resistance (R_S), also know as effective series resistance (ESR), and series inductance (L_S), also know as effective series impedance.

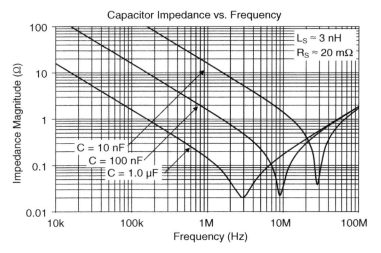

Figure 1.17: The frequency response of a capacitor varies at lower frequencies due to the series resistance and higher frequencies due to the series inductor.

2. *Know the general behavior of basic building blocks.* Consider these basic circuit cells as instruction codes. Start by using them in their most common circuit configurations or the classical approach. In analog, your basic building blocks are:

    – Analog-to-digital converters (Chapters 2 and 3)

    – Operational amplifiers (Chapters 5 and 6)

3. *Higher level thinking.* Are you afraid of math? Don't dwell on it at first. Concentrate on the practical side of analog applications. Learn the rules of thumb for analog. For

Chapter 1

instance, many of us, being indoctrinated in the school system background, sharpen our pencils, pull out the old calculator and grind through the trees before we have a thought about what the forest looks like. Once you step back and think about it, you will find that your detailed analysis can be way off. If your analysis is correct, it probably is only part of the picture. Here is a perfect example of what I mean.

Problem:

What is the corner frequency of the single-pole, low-pass R|C filter shown in Figure 1.18?

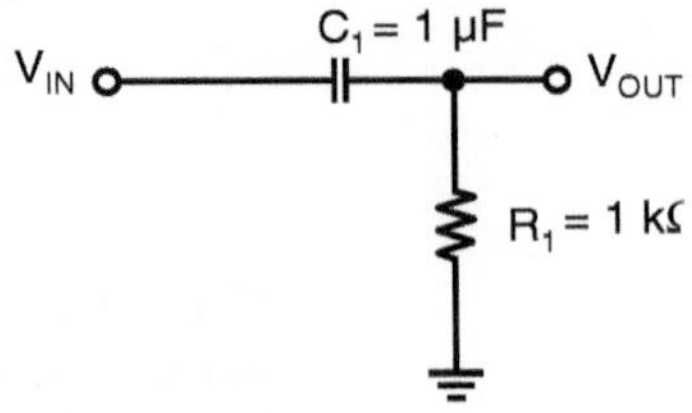

Answer:

"Hand wave" solution: Wait a minute. This isn't a low-pass filter. This is a high-pass filter. (You probably knew this right away but you would be amazed at how many would overlook this simple conclusion!) But if I assumed that the author made a mistake and reversed the placement of the resistor and capacitor, the corner frequency would be about $1/(2 \pi R \times C)$ or 160 Hz. How did I get there? Isn't $1/2\pi$ equal to about 0.16? As a first pass, I think I can accept that error because the capacitor device-to-device error is probably ±10 or 20% accurate.

Figure 1.18: Circuit example.

Calculated solution (with blinders on):

$$(V_{OUT} - V_{IN}) / (1/sC_1) = V_{OUT} / R_1$$

$$V_{OUT} (sC_1 + 1/R_1) = V_{IN} / (1/sC_1)$$

$$V_{OUT} / V_{IN} = (sR_1 \times C_1 + 1)/(sR_1 \times C_1)$$

From this calculation, there is a pole at DC and a zero at 159.1549 Hz.

These two solutions don't agree! And I bet a SPICE simulation would match your calculated solution. The moral to this story is "hand wave," or think yourself through the problem first. SPICE does not mean "don't think," it means "verification of your analysis." With this type of analysis, you should keep in mind the accuracy (or lack there of) of the various components and devices in your system. After, and only after, you know generally how the circuit works and how the system responds, give your mathematical and SPICE skills a try (see Chapter 7).

Time and Its Inversion

In the digital domain, particularly with real time operating system (RTOS), you will find that you are counting minutes, seconds, milliseconds, and nanoseconds. This is also done with analog circuits, but more importantly, the inverse of seconds is counted. Taking the inverse of seconds helps you think in terms of frequency instead of time. Frequency information is much more critical here.

4. *This could be a good career decision.* The universities are graduating less and less engineers knowledgeable in analog, but as we all know, analog will not be going away any time soon.

Organizing Your Toolbox

You need to decide what is important and what is not for your future analog design work. An effective way to do this is to arm yourself with basic, key tools of the trade. You should concentrate as you collect your ammunition on six topics.

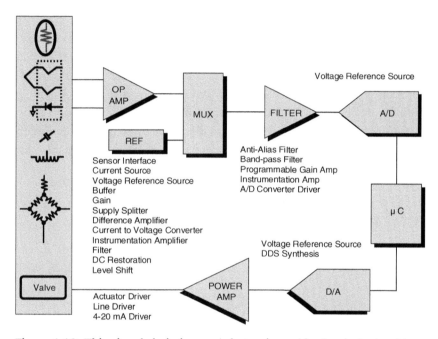

Figure 1.19: This signal chain is somewhat universal in that it deals with the analog signal coming in, conditions it through the amplification system and digitizes it in preparation for the microcontroller or processor.

First, know how to get data in and out of the digital domain. When this is mastered you will know the different topologies, important specifications, and the art of matching the converter to the application.

Then, sit back and ask yourself, "Where does my data in the controller really come from?" You will usually find some sort of sensor at the origin of the signal path. Further back from the A/D converter is the amplification system. In this system, the signal can either be enhanced through amplification or corrupted because of noise or linearity errors. The key player in the amplification system is the operational amplifier. Volumes of books have been written on this seemingly simple part, but not enough written about the single-supply

Chapter 1

operational amplifier applied in a simple manner. We will cover that in this book and take it one step further, into the battery-powered environment.

Now go back to your strength. Revisit the digital with analog in mind. Can you exploit your digital engine easily with a few analog tricks?

Go out on a limb. Bring the "art" of some of the essential analog disciplines into your toolbox. In particular, learn about noise sources and noise filters. Think about your layout and how it affects your circuit solution. Then go to the lab with confidence.

Set Your Foundation and Move On, Out of the Box

Drop your inhibition. Have fun. Work outside your box. Learning a new craft takes persistence, time and a learning attitude. Analog design is a matter of sitting down and doing it, whether it is right or wrong. Then on the next day tweak it, and the next day, and the next day, until the circuit is finally refined. No magic formulas here, just some common sense, and problem solving techniques. First, define the problem. Then identify tools and strategies that can be used to work the problem. Third, work the problem to a solution. Finally, reread your definition of the problem and determine if your solution seems reasonable. Analog only demands good, honest, consistent and persistent work. Sound familiar?

Chapter 1 References

"FilterPro™ .MFB and Sallen-Key Low-Pass Filter Design Program," Bishop, Trump, Stitt, SBFA001A, Texas Instruments.

FilterLab®2.0 User's Guide, DS51419A, Microchip Technology.

"CANbus Networks Break Into Mainstream Use", Marsh, David, EDN, Aug. 22, 2002

"Making the CANbus a "can-do" Bus," Warner, Will, EDN, Aug. 21, 2003.

"Implementing Ohmmeter/Temperature Sensor," Cox, Doug, AN512, Microchip Technology.

"Resistance and Capacitance Meter Using a PIC16C622," Richey, Rodger, AN611, Microchip Technology.

CHAPTER 2

The Basics Behind Analog-to-Digital Converters

CHAPTER 2

The Basics Behind Analog-to-Digital Converters

The analog-to-digital converter (ADC) is always in the back seat of the station wagon, looking at the analog signal through the rear window. In a way, I am soft on this device because this is where I was in my family's station wagon throughout my childhood, being one of six children. The controller, in the front seat, can see the results of the converter's labor, but the question is, can those results be counted on? If the ADC reports the system data incorrectly, the controller is blind to errors that have been introduced by the converter and signal chain. This is true unless you are willing to allocate a lot of code to try to unscramble the mess (with no guarantee of success). But why not go to the source of the problem. Believe me when I say the ADC can cause you a great deal of heartache if you don't understand the nuances. Your misunderstanding of how to use the ADC can leave the controller or processor struggling with erroneous or inaccurate data.

In this chapter, we are going to discuss the key specifications for ADCs and how they can impact your expected results from your converter. This list of specifications generally applies to all classes of converters. Then we will delve into the particulars of the successive approximation register (SAR) ADC. This part of the discussion will start with an explanation about how the SAR converter works. The issues discussed will give you insight on how to use this type of converter effectively, the first time. There will be more performance specifications and characteristics discussed here with emphasis on how to design with or around some of the converter's shortcomings. This is followed with a user-friendly version of how a sigma-delta (Σ–Δ) ADC works. The Σ–Δ topics will follow the same line of discussion as with the SAR converter. First, we will talk about the topology and in particular how it impacts your signal chain. Following this brief discussion, the performance specifications that are particular to the Σ–Δ converter will be discussed, with solutions on how to work with or work around the Σ–Δ converter limitations.

The primary ADC specifications are summarized in Appendix A and B, so if you forget about the particulars of a specification, this is a great place to look. Appendix A contains a glossary of common converter specifications. In Appendix B, you will find out what fast Fourier transform (FFT) is and how it relates to the performance of your converter.

There are numerous other converters that you can use for your application circuits, like the Pipeline, FLASH, and voltage-to-frequency (V/F) converters, but these topologies are beyond the scope of this book.

Chapter 2

The Key Specifications of Your ADC

Input Range of the ADC

The input range of the ADC can be a bit tricky. You will find variations of single-ended, differential, pseudo-differential, while the input range is determined by the voltage reference (V_{REF}) of the converter.

An example of the configuration of an ADC with a single-ended input is shown in Figure 2.1a. This type of converter input is easy to use because there is no question of what to do with that pin. The input voltage range is equal to the full-scale range (FSR) of the converter. Additionally, the digital code at the output of this configuration is straight binary (see the Straight Binary Section in this chapter).

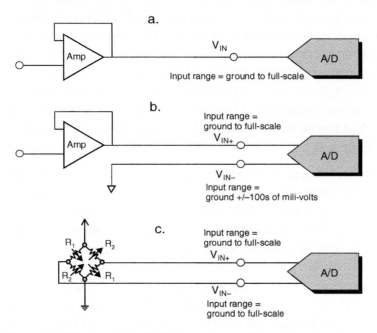

Figure 2.1: The input(s) of ADCs can be configured in one of three ways. The single-ended input (A) is configured for one input voltage referenced to ground. Another type of input stage has two inputs configured as a "pseudo-differential" stage (B) where the signal input is the noninverting input, and the inverting input is used to reject small-signal system noise. The third type of input stage is the differential input (C) where the two inputs to the converter range from ground to the full-scale input voltage.

In Figure 2.1b, the input of the converter is configured as a pseudo-differential input. This simply means that the input to the converter is differential, but one of the input pins has a range that is limited to a few hundred millivolts from ground. This has the same output digital coding as the single-ended input device. The digital code at the output is straight binary. You might ask what this configuration would do for you? True to the spirit of differential input stages, this type of device will reject small common-mode noise. In simple terms, if a small-signal, like 50 Hz or 60 Hz, is an undesirable part of the signal you are trying to convert, this common-mode signal will be rejected or eliminated. This is a nice feature as long as you understand that you have to connect it properly to ground.

A third type of input for ADCs is the fully differential input stage. With this configuration both inputs can be brought from ground to the converter's FSR. This is nice, because not only can you reject small common-mode signals but you also can convert a positive or negative analog signal to a digital output. You guessed it, the output code is in the format of two's complement (see the Two's Complement section in this chapter).

So why is this important? There are some signal sources that are differential. One example is the signal from a Wheatstone bridge in Figure 2.1c. You will find that if one of the inputs goes positive, the other will go negative. This action will give you a gain of 2 V/V in the analog domain. It will also couple the noise in the environment on both lines. The ADC will then be able to filter that noise and convert the signal of interest from the bridge.

With this type of input, the FSR is double what you might expect. If you think about it, if the noninverting input goes to its full range and the inverting input is referenced to ground, the difference in this signal is $(V_{IN+} - V_{IN-}) = +V_{FS}$. Now if the input signal changes so that the noninverting input is referenced to ground, and the inverting input is taken to full-scale, the difference in the input signal is $(V_{IN+} - V_{IN-}) = -V_{FS}$. So the actual FSR of this type of device is $+V_{FS} - (-V_{FS})$ or $2 \times V_{FS}$, where V_{FS} is the full-scale input voltage range.

Digital Coding of the Analog Signal

An analog-to-digital converter translates an analog input signal into a discrete digital code. This digital representation of the *real-world* signal can be manipulated in the digital domain for the purposes of information processing, computing, data transmission, or control system implementation. In any application where a converter is used, it is advantageous to have the code structure complement the microcontroller or processor's operands.

We are going to talk about the most common code schemes: Straight binary and binary two's complement code. For simplicity, all of the following code examples are for a 4-bit converter. The median analog voltages in the tables are the equivalent analog voltages that are at the center of the digital code.

These codes are mathematically described using the full-scale (FS) input range of the converter. Usually the full-scale input range of an ADC is equal to or twice as much as the voltage reference applied to the device. And in some instances, the voltage reference connect is tied internally in the device to the power supply. In all of these configurations, you will need to refer to the product's data sheet for specifics.

Chapter 2

The basic differences between these two types of code is:

1. The least significant bit (LSB) size of the two's complement code is twice as large as the LSB size of the straight binary code. This does not increase the number of codes that the converter can create. With the two's complement code, the positive and negative voltage inputs to the converter are represented with the same amount of codes that the straight binary represents, only the positive voltage input voltage with code.

2. The analog FSR of the straight binary code is a positive voltage from ground to V_{REF}. The analog FSR of the two's complement code is equal to the positive FSR, plus the unsigned negative FSR.

3. The digital output code of the two's complement code is easier when running arithmetic calculations, such as subtraction.

Straight Binary Code

The straight binary code is more accurately called unipolar straight binary. This digital format for an analog-to-digital conversion is the simplest to understand. As the name implies, this coding scheme is used only when positive voltages are converted. This is a good output code for converters that are configured with a signal input, as shown in Figures 2.1a and 2.1b. An example of this type of coding is shown in Table 2.1.

Table 2.1: The Unipolar Straight Binary Code representation of zero volts is equal to a digital 0000. The analog full-scale minus one LSB digital representation is equal to 1111. With this code there is no digital representation for analog full-scale.

Median analog voltage (V)	Digital code
0.9375 FS ($^{15}/_{16}$ FS)	1111
0.875 FS ($^{14}/_{16}$ FS)	1110
0.8125 FS ($^{13}/_{16}$ FS)	1101
0.75 FS ($^{12}/_{16}$ FS)	1100
0.6875 FS ($^{11}/_{16}$ FS)	1011
0.625 FS ($^{10}/_{16}$ FS)	1010
0.5625 FS ($^{9}/_{16}$ FS)	1001
0.5 FS ($^{8}/_{16}$ FS)	1000
0.4375 FS ($^{7}/_{16}$ FS)	0111
0.375 FS ($^{6}/_{16}$ FS)	0110
0.3125 FS ($^{5}/_{16}$ FS)	0101
0.25 FS ($^{4}/_{16}$ FS)	0100
0.75 FS ($^{3}/_{16}$ FS)	0011
0.1875 FS ($^{2}/_{16}$ FS)	0010
0.0625 FS ($^{1}/_{16}$ FS)	0001
0	0000

When this scheme is used to represent a positive analog signal range, the digital code for zero volts is equal to zero (0000 per Table 2.1). The definition of a positive voltage is the amplitude between the ADC ground or the inverting input and the noninverting input of the ADC. Given an ideal converter with no offset, gain, integral nonlinearity (INL), differential nonlinearity (DNL) error or noise, the code transition from 0000 to 0001 occurs at the analog value of:

First Code Transition = $(0 + \frac{1}{2}$ LSB$)$

Second Code Transition = $(1$ LSB $+ \frac{1}{2}$ LSB$)$

where

LSB = +FS / 2^n

Where LSB is equal to the least significant bit

n is equal to the number of converter bits and

+FS is equal to the analog FSR.

The digital output code versus analog input voltage is mapped in Figure 2.2. In this figure, the analog input voltage of 0 V is converted to a digital code of 0000. If the analog input voltage is changed at the input of the ADC to approximately 0.25 FS voltage, the ADC will produce a digital code of 0100. At approximately midpoint between the codes specified in Table 2.1, the converter will transition from a lower digital code to a higher digital code. A code width as defined in Figure 2.2 is equal to one least significant bit (LSB).

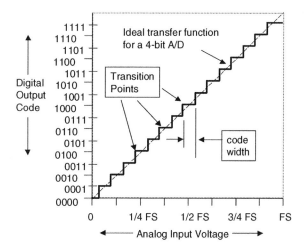

Figure 2.2: This is an ideal transfer function of a 4-bit ADC where the analog input is continuous and the digital output code is discrete.

Binary Two's Complement Code

In some applications, the unipolar ADC converts negative and positive values. This can only happen when the ADC has differential analog inputs, as shown in Figure 2.1c. These types of devices will output digital code in the binary two's complement format. Binary two's complement arithmetic is widely used in microcontrollers, calculators and computers because simple subtractions and additions require less code. The additional operation of two binary numbers

Chapter 2

in two's complement is straightforward, in that the two numbers are added together. The subtraction operation of numbers in two's complement is done by adding the two numbers together to get the subtracted solution.

For example:

Decimal	Two's Complement	Decimal	Two's Complement	Decimal	Two's Complement
+7	0111	+7	0111	+5	0101
+5	0101	−5	1011	−7	1001
+12	1100	+2	0010	−2	1100

Binary two's complement coding is not as straightforward as straight binary. The codes are not continuous from one end to the other due to the discontinuity that occurs at the analog bipolar zero. The two's complement representation of a positive binary number is generated by logically complementing all the digits, which then converts it to the negative binary number counterpart as shown in Table 2.2 and Figure 2.3. With this code scheme, the most significant bit (MSB) is a sign indicator. A positive value is indicated with an MSB logic 0. An MSB value of logic 1 indicates that the output number is a negative value.

Table 2.2: The binary two's complement representation of zero volts is also equal to a digital 0000. The analog positive FS minus one LSB digital representation is equal to 0111, and the analog negative FS representation is 1000.

Median Voltage (V)	Code
0.875 FS ($^7/_8$ FS)	0111
0.75 FS ($^6/_8$ FS)	0110
0.625 FS ($^5/_8$ FS)	0101
0.5 FS ($^4/_8$ FS)	0100
0.375 FS ($^3/_8$ FS)	0011
0.25 FS ($^2/_8$ FS)	0010
0.125 FS ($^1/_8$ FS)	0001
0	0000
−0.125 FS (−$^1/_8$ FS)	1111
−0.25 FS (−$^2/_8$ FS)	1110
−0.375 FS (−$^3/_8$ FS)	1101
−0.5 FS (−$^4/_8$ FS)	1100
−0.625 FS (−$^5/_8$ FS)	1011
−0.75 FS (−$^6/_8$ FS)	1010
−0.875 FS (−$^7/_8$ FS)	1001
−1 FS	1000

The Basics Behind Analog-to-Digital Converters

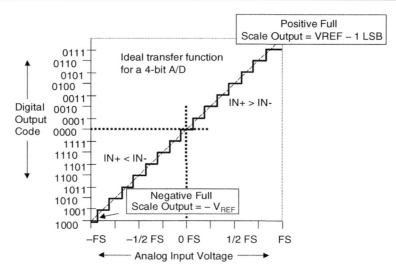

Figure 2.3: The transfer function of a 4-bit ADC bipolar analog input will produce a two's complement code as a digital output.

This system has an odd number of codes and only one zero state. Differential input ADCs (Figure 2.1c) are devices that can be operated in a single-ended, positive voltage input mode or a full-differential input mode. In the full-differential mode, the FSR of the device is equal to:

$$FSR = \{+IN_{MAX} - (-IN_{MIN})\} + \{-IN_{MAX} - (+IN_{MIN})\}$$

And the ADC's input voltage range is equal to:

$$A_{IN} = (+IN - (-IN))$$

The converter will produce digital code that represents the both negative and positive analog inputs as shown in Table 2.2.

Throughput Rate versus Resolution and Accuracy

I once had a customer ask me on the company hotline for a 32-bit converter. I was taken off guard with his request. Why did he need 32-bits? I finally found out that he only needed 1 mV resolution out of 4.096 V. Well that is easy. You can do this with a 12-bit converter that has a 4.096 V reference. With this converter, the FSR would be 4.096 and the LSB size would be $V_{REF} / 2^{12} = 4.096$ V/4096 = 1 mV. Why was he having problems finding a converter?

He told me that his bus was 32-lines wide. So my advice became, "Use a 12-bit converter and tie bus lines 0 through 29 to ground." He asked me if you could do that and I said, "Absolutely, and with this solution you will control your system noise as an added benefit!"

A *throughput rate* specification (also know as data rate for sigma-delta ADCs) defines that amount of time it takes for a converter to complete an entire conversion. The activities that

Chapter 2

are included in the throughput rate time are setup time, sample time, conversion time and data transmission time. The two converter topologies that we will center our attention on will be the successive approximation register (SAR) and sigma-delta. When you think of the typical conversion times of these topologies, you can easily separate them into application classes (Figure 2.4). In general terms, the faster the throughput rate of the converter, the lower the resolution will be. But implied in this diagram is the same trend in accuracy. The resolution of a converter is simply the number of bits that the converter is capable of handling per conversion. The accuracy describes the number of bits that are repeatable from conversion to conversion. In all cases, the specified accuracy will be equal to or less than the resolution of the converter.

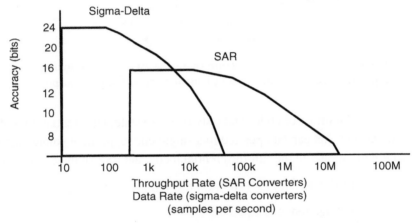

Figure 2.4: The throughput rate of the SAR converter is faster than a sigma-delta converter. In contrast, the sigma-delta converter is able to achieve higher accuracy as a trade-off for the slower speeds.

The data rate of the sigma-delta converter is generally slower than the throughput rate of the SAR converter. As we will see later on in this discussion, the SAR converter only samples the input signal once and converts to a digital code, according to the sampled signal. The sigma-delta converter samples the input signal multiple times. It then implements various noise reduction algorithms to improve the number of bits in the converter, as well as the signal-to-noise ratio (SNR), but the trade-off with this type of sampling strategy is time.

Accuracy versus Resolution

There are a few key specifications that you should become familiar with. Knowing these figures of merit will help you choose the right converter for your application and also identifies the impostors. For instance, "2.7V 16-Bit ADC with SPI® Serial Interface" is an example of what you might see at the top of a data sheet. Does it mean 16-bits accurate, noise-free for every conversion, accurate with respect to the input voltage, or does it mean 16-bits resolution where you are guaranteed that 16-bits will be transmitted out of the converter at the conclusion of a conversion? The latter is correct. The phrase "2.7 V 16-Bit ADC with SPI Serial Interface" as the title of the converter's data sheet only means that you will see 16-bits transmitted from the output of the converter.

You will find that those 16-bits can all be accurate or not, depending on the manufacturer. More than once in my career I have seen the last couple of codes or the LSBs (least significant bits) of a 16-bit converter dither all over the place from conversion to conversion. So, I would define a converter like this as having 16-bit resolution, not 16-bit accurate. This is not a bad thing, as long as you know what to expect.

So, resolution is defined as the number of bits that are transmitted out of the converter at the conclusion of the conversion. If you know this information about a converter you can quickly calculate the theoretical LSB size with the following formula:

$$LSB = FSR / 2^n$$

Where n = number of bits

For a converter that has 16-bit resolution and an FSR of 5 V, the LSB size is 76.29 µV.

One of the more common questions about ADCs that I hear is, "How do I know that an ADC will give me a good, reliable code and can I determine this from the converter's data sheet?" Of course this depends on your definition of "reliable," but if you are looking for a repeatable output from conversion-to-conversion, you should refer to AC domain specifications. If you are looking for a converted code that represents that actual input voltage, DC specifications are more useful. But, don't forget about the noise. DC specifications imply average accuracy (not repeatability). From conversion-to-conversion these codes will vary, dependent on the internal noise of the converter.

AC Specifications Imply Repeatability

AC domain specifications, such as SNR, effective resolution (ER), signal-to-(noise + distortion) (SINAD), or effective number of bits (ENOB), provide information about ADC repeatability. Now these specifications will tell you how repeatable your conversion is, but they will not tell you if the conversion is accurate. On the other hand, DC domain specifications, such as offset error, gain error, differential nonlinearity and integral nonlinearity provide information about how close, on average, the input signal is matched to an actual output code. These specifications do not imply repeatability, and noise could give you varying results from conversion-to-conversion.

Ideally, the SNR of a converter in decibels is equal to 6.02 n + 1.76 dB, where "n" is equal to the number of converter bits. This theoretical noise is a result of the quantization noise inherent in the converter. In practice, SNR is equal to 20 log (rms signal)/(rms noise), where rms

Chapter 2

means root-mean-square, equal to one standard deviation in a normal distribution. In order to determine the rms noise, the results of many conversions need to be collected.

As with the SNR, ER is measured by collecting a statistical sample of many conversions, but this time we don't have an AC input signal to the converter. The input signal is a clean, "noiseless" DC signal. If this DC signal has less noise than your converter (about 3×), you are good to go. The units of measure for ER are bits, which is referred to the output of the converter. If you want to refer this number to the input, you can change the units to volts. This is interchangeable with the following formula:

$$\text{ER (in bits rms)} = \{20 \log (\text{FS} / \text{ER in V}_{rms}) - 1.76\}/6.02,$$

since the unit of ER is in bits, you now know which bits are repeatable in your ADC output code.

While SNR or ER provides information about the device noise of the converter, SINAD and ENOB provide more information about ADC frequency distortions. SINAD is the ratio of the rms amplitude of the fundamental input frequency of the input signal to the rms sum of all other spectral components below one half of the sampling frequency (excluding DC). The theoretical minimum for SINAD is equal to the SNR or $6.02 n + 1.76$ dB. But in practice, an ADC will have some harmonic distortion of this input signal that is generated within the converter. The complementary specification to SINAD is ENOB. The unit of measure for SINAD is dB, and the units of measure for ENOBs is bits. SINAD can be converted to ENOB with the following calculation:

$$\text{ENOB} = (\text{SINAD} - 1.76)/6.02$$

To this point in our discussion, the specification units are in terms of rms. Statistically speaking, rms to one standard deviation of data shaped as a normal distribution. When the noise units are defined with rms units, the probability that the converter will give you a value of plus or minus one rms is ~68%. The relationship between the output noise of the converter, the normal distribution of a set of sampled outputs and these statistical values are illustrated in Figure 2.5a and 2.5b.

An rms specification is a statistical calculation from many samples or a population. The formula for one standard deviation is:

$$\sigma^2 = \Sigma (y - \eta)^2/N$$

where σ is the population standard deviation

y is a sample from the population

η is the population mean and

N is the set of population observations

The Basics Behind Analog-to-Digital Converters

A 68% probability of getting your expected output may not be the odds you want to work with. You might want to consider converting the specification limits to peak-to-peak (p-p) values. From the rms number, you can quickly calculate the p-p specification, which is very convenient if you are trying to get good repeatable results. This conversion is easily done with ER and ENOB specifications by multiplying your rms specification (in voltages) by two times the crest factor (CF, Figure 2.5c), or subtracting the bit crest factor (BCF, Figure 2.5c) from your rms specification (in bits). With this new calculation, your ADC has a better chance of producing your expected output. The industry standard crest factor for nonmilitary applications is 3.3. Also, be careful that the data you selected has the attributes of a normal distribution, otherwise these calculations will not be as accurate as promised.

The calculation for the conversion of rms to p-p is:

V(p-p) = V(rms) * 2 * CF

or

Bits (p-p) = Bits (rms) − BCF

For more details about these specifications, refer to Appendix A.

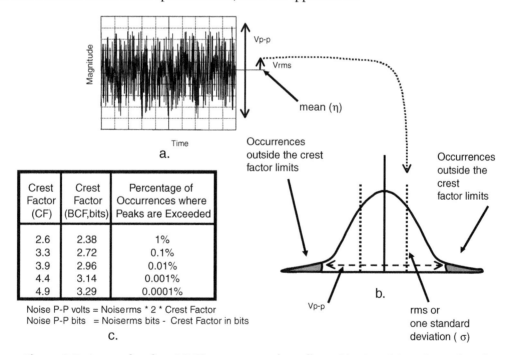

Figure 2.5: A sample of an ADC's output can be collected in time (a) and translated into a histogram (b) where the mean and standard deviation of the samples can be calculated. With the standard deviation of these samples, a peak-to-peak value can be determined (c) with a multiple of 2 times the crest factor (CF) for output signal referred to input calculations or an additive bit crest factor (BCF).

Chapter 2

DC Specifications Imply Accuracy

If you are looking for a conversion from your converter that accurately represents the analog voltage, you should be looking at the DC specifications. The DC specifications that I am referring to are offset voltage, gain error (or FS error), differential nonlinearity, and integral nonlinearity. If your conversions are not repeatable as discussed in the "AC Specifications Imply Repeatability" section in this chapter, then the accuracy of your converter is determined by the average of multiple samples. In this discussion, we are going to assume that the converter is noise-free.

What do the least significant bit (LSB) specifications mean when you are looking at ADCs? One day a fellow engineer told me that a 12-bit, Converter X (manufacturer will remain unnamed) had just 7 usable bits. So essentially, the 12-bit converter was only a 7-bit converter. He based this conclusion on the device's offset and gain specifications. The maximum specifications were:

$$\text{Offset error} = \pm 3 \text{ LSB},$$
$$\text{Gain error} = \pm 5 \text{ LSB},$$

At first glance, I thought he was right. From the list above, the worst specification is gain error (±5 LSB). Applying simple mathematics, 12-bits of resolution minus five is equal to 7-bits, right? Why would an ADC manufacturer introduce such a device? The gain error specification motivates me to purchase a lower-cost, 8-bit converter. However, that doesn't seem right. Well, as it turned out, it wasn't right.

Let's start out by looking at the definition of LSB. Think of a serial 12-bit converter; it produces a string of twelve ones or zeros. Typically, the converter's first transmitted digital bit is the MSB (or LSB + 11). Some converters transmit the LSB first. We will assume that the MSB is first in this chapter. The second bit is MSB − 1 (or LSB + 10); the third bit is MSB − 2 (or LSB + 9), and so on. At the end of this string of bits, the converter finally transmits as MSB − 11 (or LSB).

The terminology, LSB, is very specific. It describes the last position in the digital stream. It also represents a fraction of the full-scale input range. For a 12-bit converter, the LSB value is equivalent to the analog full-scale input range divided by 2^{12} or 4096. If I put this in terms of real numbers, I have an LSB size of 1 mV with a 12-bit converter that has a full-scale input range of 4.096 V. However, the most instructive definition of LSB is that it can represent one code out of the 4096 codes possible.

Going back to the specifications and translating them into a 12-bit converter that has an input FSR of 4.096 V:

$$\text{Offset error} = \pm 3 \text{ LSB} = \pm 3 \text{ mV},$$
$$\text{Gain error} = \pm 5 \text{ LSB} = \pm 5 \text{ mV},$$

These specifications actually claim that the converter can have (worst case) an 8 mV (or 8 code) error introduced through the conversion process. This is not to say that the error occurs at the LSB, LSB – 1, LSB – 2, LSB – 3, LSB – 4, LSB – 5, LSB – 6 and LSB – 7 positions in the output bit stream of the converter. The errors can be up to eight times one LSB, or 8 mV. Precisely stated, the transfer function of the converter could have up to eight codes missing out of 4096 codes. These codes will be missing at the lower or upper range of the codes. For example, a converter with an error of +8 LSB ((+3 LSB offset error) + (+5 LSB gain error)) will produce possible output codes of zero to 4088. The lost codes are from 4088 up to 4095. This is a small, incremental error of 0.2% at full-scale. In contrast, a converter with an error of –3 LSB ((–3 LSB offset error) – (–5 LSB gain error)) will produce codes from 3 up to 4095. The gain error in this situation produces an accuracy problem, not a loss of codes. The lost codes are 0, 1, and 2. Both of these examples illustrate the worst possible scenario.

The difference between the first measured transition point and the first ideal transition point is the offset voltage of the converter. If the offset error is known it can easily be calibrated out of the conversion in hardware or software by subtracting the offset from every code. Gain error (full-scale error) is the difference between the ideal slope from zero to FS and the actual slope between the measured zero point and FS. Offset errors are zeroed out with this error calculation. Gain error is another ADC characteristic that can be calibrated out of the final digital code from the converter. Multiplying the final conversion by a constant does this. Although this calibration is possible, the software overhead may be too much. Typically, the offset errors and gain errors do not track this closely in actual converters.

The real-life performance enhancements due to incremental improvements in an ADC's offset or gain specifications are negligible to nonexistent. To some designers, this seems like a bold assumption, if precision is one of the design objectives. It is easy to implement digital calibration algorithm with your firmware. However, more importantly, the front-end amplification/signal conditioning section of the circuit typically produces higher errors than the converter itself.

This discussion puts a new light on the conclusions reached at the beginning of this section. In fact, the 12-bit converter as specified above has an accuracy of approximately 11.997 bits. The good news is that a microprocessor or microcontroller can remove this offset and gain error with a simple calibration algorithm.

Differential nonlinearity (DNL) is the maximum deviation in code width from the ideal 1 LSB ($FS/2^n$) code width. The difference is calculated for each transition. This converter characteristic is very difficult to calibrate out. Even if you take the time to measure one converter for this error, the next converter from the same product family will have a slightly different DNL error from code to code. Integral nonlinearity (INL) is the maximum deviation of a transition point from the corresponding point of the ideal transfer curve with offset and gain errors zeroed. The INL performance of an ADC is actually derived from DNL tests.

Once again, the INL error is difficult to calibrate out of the final conversion, particularly from part-to-part of a product family.

For more details about these specifications, see Appendix A.

Successive Approximation Register (SAR) Converters

The SAR ADC arose out of industrial application requirements. This tried but true converter solution has spread across a variety of applications including process control, medical and earlier audio systems. In these applications, 8- to 16-bit conversion results were required.

The SAR ADC is nothing new to the data acquisition world. In the 1970s, the state-of-the-art SAR ADC was touted as a lower power, more accurate and less expensive device. These converters utilized R-2R resistive ladders in their design in order to achieve the differential linearity, integral linearity, offset and gain specifications. They were able to achieve the promised performance because of careful IC layout practices and wafer level resistor laser trimming. The core of this first generation SAR ADCs required an external sample-and-hold circuit but was exclusively built using a bipolar transistor process. This was a good marriage because the bipolar technology was best suited for low-noise and high-speed performance. A good example of this type of converter would be the industry standard, ADC700, manufactured by Texas Instruments.

In today's standards, this hybrid ADC would be considered too power hungry. The current CMOS generation of SARs has succeeded in taking over the all-bipolar SAR. The architecture of this converter uses a capacitive redistribution input section, which inherently includes the sample/hold function. The capacitor arrays are more compact and much easier to match than the older nicrome R-2R ladder networks, which usually requires an external sample-and-hold circuit on the analog front end. This new chip topology has lower power operation, higher functionality and a smaller chip size.

All this is good news to the systems designer who is looking for improved performance, higher integration and an overall excellent cost/performance ratio. This generation of SAR converters not only includes the sample-hold function, but also differential inputs and voltage-controlled gain capability through the voltage reference inputs. Since the integrated circuit design is implemented primarily with capacitors rather than resistors, the power dissipation and the chip size is lower than ever achieved. The SAR converter also has taken a step towards increased functionality. In prior SAR ADC designs, the voltage reference circuit could be internal or external, but in all cases was limited in voltage range. With this new topology, the device voltage reference is usually external and its range is much wider. This gives flexibility when selecting the desired LSB size. As mentioned before, the LSB size of a converter is:

$LSB = FSR / 2^n$
where n = number of bits,
and FSR = the voltage reference voltage.

Under normal single-supply conditions, the voltage reference would be equal to 5 V. If this is the case, the LSB size of a 12-bit converter is equal to 1.22 mV (5 V ÷ 4096 codes). If the voltage reference for the converter is equal to 100 mV, the LSB size now becomes 0.0244 mV. This is a 50× reduction in the LSB size. If you have a very clean layout and voltage reference, this type of change could eliminate an analog gain stage.

A final advantage of the CMOS version of the SAR converter is that it is possible to integrate this circuit onto the microcontroller or processor chip. This is not feasible with the bipolar SAR converter unless you produced an expensive multichip, mixed-signal version.

The CMOS SAR Topology

The CMOS SAR ADC is a sampling system that takes one sample for every conversion. The analog input signal to a SAR converter first sees a switch and a capacitive array, as shown in Figure 2.6. The input node connects a capacitive array on one side, and the noninverting input to a comparator on the other.

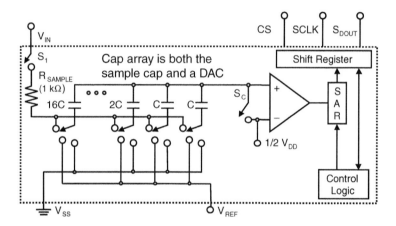

Figure 2.6: The modern day SAR converter uses a capacitive array at the analog input. This capacitive array and the remainder of the device are easily manufactured in CMOS, making it easy to integrate it with microcontrollers or microprocessors.

Chapter 2

When the switch (S_1) is closed, the voltage input signal is sampled onto the internal capacitive array of the converter. After the sampling time is completed, S_1 is opened and the bottom side of the MSB (most significant bit) capacitor is connected to V_{REF} while the other capacitors are tied to V_{SS} (or the system ground). The charge from the MSB capacitor is redistributed among the other capacitors. The charge is distributed across the capacitor array, and the noninverting input of the comparator moves up or down according to the voltage presented at its input. The voltage at the noninverting input of the comparator, with respect to V_{SS}, is equal to $(1/2V_{DD} - V_{IN}) + 1/2V_{REF}$. If this voltage is greater than $1/2V_{DD}$, the MSB is equal to zero, which is transmitted out of the serial port, and the MSB capacitor is left tied to V_{REF}. The transmission of all bits to the serial port is synchronized with system clock (SCLK) through S_{DOUT} (serial data out). If the voltage at the noninverting input of the comparator is less than $1/2V_{DD}$, the MSB capacitor is connected to V_{SS} and an MSB bit equal to one is transmitted out of the serial port.

As soon as the MSB value is determined, the converter starts to determine the MSB – 1 value. Connecting the MSB – 1 capacitor to V_{REF} while the other capacitors are tied to V_{SS} (except for the MSB capacitor) does this. Note that the MSB – 1 capacitor is not illustrated in Figure 2.6, but its value is 8C. With this change in the capacitive array connections, the value of the voltage at the noninverting input of the comparator is $[1/2 V_{DD} - V_{IN}] + 1/2V_{REF} (MSB) + 1/4V_{REF}$. Now the voltage on the capacitive array is compared to the voltage at the inverting of the input comparator, $1/2V_{DD}$. In the analysis of this bit, if this voltage is greater than $1/2V_{DD}$, the MSB – 1 is equal to zero, which is transmitted out of the serial port. Additionally, the MSB – 1 capacitor is left tied to V_{REF}. If the voltage across the capacitive array is less than $1/2V_{DD}$, the MSB – 1 bit is equal to one. This bit value is transmitted out through the serial port. Once this is done the MSB – 1 capacitor is connected to V_{SS}. This process is repeated until the capacitive array is fully utilized.

There are two critical points during the conversion time. The first point is where the sample is actually acquired by the converter. During this time, the input signal must be stable within ¼ of an LSB. Otherwise, the converter will give an output that is less than accurate. The second critical point during the conversion time is where the converter is finishing up the conversion. At this particular time, the converter is converting the LSB, which requires the most accuracy. Generally speaking, it is good practice to keep the converter's power supply and input signal as quiet as possible during the entire conversion.

Figure 2.7 shows another way of thinking about the SAR conversion by looking at the digital-to-analog converter (DAC) output. In this figure, the input is sampled between time (a) and (b). Starting at time (b), the analog voltage is tested against the DAC output voltage, which is now equal to ½ FS. If the analog input voltage is higher than ½ FS, a digital output code of "1" is sent out of the serial digital output. If the charge from the analog input voltage is lower than ½ FS, a digital output code of "0" is sent out of the serial digital output. In this case, the MSB value is "1". The capacitive array is switched to test MSB – 1 as discussed above.

The Basics Behind Analog-to-Digital Converters

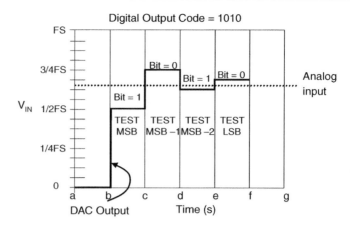

Figure 2.7: Another way of looking at the SAR conversion process is to examine the internal DAC output of the converter. The converter starts by converting the MSB of the analog input signal and then steps through each bit. Each bit conversion is timed with the system clock.

Between time (c) and (d) the analog input charge is now compared to ¾ FS. If the MSB was found to be a "0" the MSB – 1 bit would be compared to ¼ FS. But as you can see in this case, with the MSB equal to "1", the MSB – 1 is determined to equal "0". This process continues until the final LSB code is determined.

Interfacing With the Input of the SAR Converter

Driving any A/D converter can be challenging if all issues and trade-offs are not well understood from the beginning. With the SAR converter, the sampling speed and source impedance should be taken into consideration if the device is to be fully utilized. Here we will discuss the issues that surround the SAR converter's input and conversion to ensure that the converter is handled properly from the beginning of the design phase. We are also going to review the specifications available in most A/D converter data sheets, and identify the important specifications for driving your SAR. From this discussion, techniques will be explored which can be used to successfully drive the input of the SAR A/D converter. Since most SAR applications require an active driving device at the converter's input, the final subject is to explore the impact of an operational amplifier on the analog-to-digital conversion in terms of DC as well as AC responses. A typical system block diagram of the SAR converter application is shown in Figure 2.8. Some common SAR converter systems are data acquisition systems, transducers sensing circuits, battery monitoring applications and data logging. In all of these systems, DC specifications are important. Additionally, the required conversion rate is relatively fast

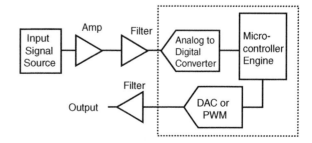

Figure 2.8: An input signal to the SAR converter should be buffered to reduce impedance matching problems and a filter to reduce aliasing errors in the converter. The amplifier stage and filter stage in this diagram can be combined.

Chapter 2

(as compared to sigma-delta converters) and having a lower number of bits that are reliably converted is acceptable.

For the input stage of the converter shown in Figure 2.9, the input signal could be AC, DC or both. The operational amplifier is used for gain, impedance isolation and its drive capability. A filter of some sort (passive or active) is needed to reduce noise and to prevent aliasing errors.

A model of the SAR ADC internal input sampling mechanism is shown in Figure 2.9. Critical values in this model are R_S, C_{SAMPLE} and R_{SWITCH}. C_{SAMPLE} is equivalent to the summation of the capacitive array shown in Figure 2.6. Pin capacitance and leakage errors are minimal. The external source resistance and sample capacitor combines with internal switch resistance and internal sample capacitor to form an R/C pair. This distributed R/C pair requires approximately 9.5 time constants to fully charge to 12-bits over temperature. The MCP3201 (from Microchip), 12-bit ADC requires 938 nsec to fully sample the input signal, assuming $R_S << R_{SWITCH}$.

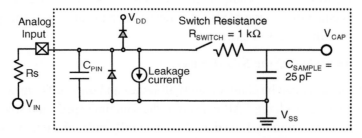

Figure 2.9: The combination of the external input resistance (R_S) and internal switch resistance (R_{SWITCH}) and the sample capacitor (C_{SAMPLE}) form a low-pass filter that has a rise time. If an accurate conversion is desired, the external input resistance should be minimized or the sampling time of the converter should be lengthened.

Following these elements, the signal reaches the switch resistance, R_{SWITCH}, and the sample capacitor, C_{SAMPLE}. The sampling capacitor represents the bulked element that samples the input signal while the switch is closed. While the converter is sampling the input signal, the combination of the source resistance (R_S), the switch resistance (R_{SWITCH}), and the sampling capacitor (C_{SAMPLE}) form a single-pole R/C network. The time constant of this network is:

$$t_{RC} = (R_S + R_{SWITCH}) * C_{SAMPLE}$$

Assuming that the charge and voltage on the sample capacitor is zero at the time that the sample is acquired, the rise time of the voltage on that capacitor is equal to:

$$V_{CAP} = V_{IN} (1 - e^{-t/(RS+RSWITCH)(CSAMPLE)})$$

The result of having a high external input resistance or fast conversion time can compromise the accuracy of the conversion. In either case, the converter will close its sampling switch before the sample capacitor has completely charged. This is illustrated in Figure 2.10.

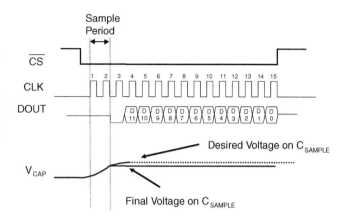

Figure 2.10: The converter is allowed to sample the input signal for a defined amount of time. The system clock to the converter determines this time. If the sampling time is too fast or the external input resistor is too high, the sample capacitor will not completely sample the input signal resulting in an inaccurate conversion.

With the previous formula, we can determine the percentage of charge that will arrive at the sample capacitor with time.

If you apply this concept to a specific application where a 12-bit A/D converter from Figure 2.9 is used, you can calculate the number of bits that you have acquired from the input signal. This is shown in Table 2.3.

Table 2.3: The R/C pair of R_{SWITCH} = 1 kΩ and C_{SAMPLE} = 25 pF requires approximately 9.5 time constants to fully charge to 12-bits over temperature.

# of time Constants	1	5	8	9	10
$(R_S + R_{SWITCH}) * C_{SAMPLE}$ in ns	25	125	200	225	250
% of Full Scale Range on C_{SAMPLE}	63.2	99.3	99.966	99.9877	99.9955
% of Full Scale Range on C_{SAMPLE} to go	36.8	0.67	0.034	0.0123	0.0045
ADC bit Accuracy (bits)	1.4	7.2	11.5	13.0	14.43

As calculated in the table, the accuracy of an analog-to-digital converter can be compromised if the device is not given enough time to sample. For instance, use the example of a fictitious 12-bit A/D converter, which samples within 1.5 clock cycles using a clock rate of 2 MHz. The sampling time allotted by the converter is 750 nsec. This works very well with the numbers in Table 2.3. Now add a source resistance of 5 kΩ and you will find that the converter needs 1350 nsec to convert to 13 bits accurately.

The accuracy of a SAR ADC can be compromised if the device is not given enough time to sample. In the graph in Figure 2.11, the y-axis is the clock frequency in megahertz, and the x-axis is input (source) resistance in ohms. The sampling time of the converter for these clock frequencies is equal to 1.5 clocks. For example, a clock speed of 1.6 MHz would translate to a sample time of (1.5/1.6 MHz) or 937.5 nsec.

Chapter 2

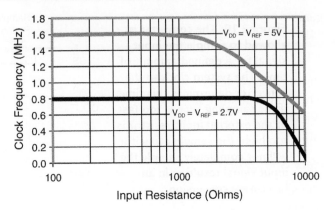

Figure 2.11: There is a level of source resistance that a SAR converter can tolerate with a given sampling frequency. But as the external source resistance increases the converter reaches a point where the sampling capacitor needs more time to charge. These variables change from converter to converter. The product data sheet from your converter should be consulted.

There are two obvious solutions to the problem: one would be to reduce the source resistance, while a second would be to increase the sampling time.

In order to keep the source resistance low, it is recommended that the converter be driven by an active element, such as an operational amplifier. In this situation, the input signal could be AC, DC or both. The operational amplifier can be used for gain, filtering, impedance isolation and its drive capability. When you drive the input of an ADC with an operational amplifier, whether it is a gain cell, filter cell or both; offset, noise, gain errors and distortion can be added to the signal prior to the ADC by the amplifier. The investigation of these issues as they relate to the conversion process follows. The selection of the appropriate operational amplifier for the SAR converter is discussed in Chapter 5, Finding the Perfect Op Amp (Precision) for Your Perfect Circuit.

Sigma-Delta (Σ–Δ) Converters

I remember the days when I could count on enough good analog circuit questions to create quite an afternoon of adventures in the lab. I would use theory to predict the analog outcome in terms of stability, gain, or noise levels and try to match it with reality. The mixed-signal ADC and digital-to-analog converter (DAC) questions were riddled with digital timing problems to conquer, but seldom the complex noise or stability calculations that pure analog circuits provided. Since the mixed-signal circuits still required an analog front-end (gain and filtering), most calculations were still handled in the analog domain. The entrance of the 16-bit converter offered somewhat of a front-end noise reduction challenge, but it did not rival the battle being waged in the pure analog domain.

Then a new player joined the team. This player was literally thrown over the wall, at least for those who aren't IEEE fans. I couldn't imagine replacing these complex circuits with a digital-centric device, but I should have paid attention to the reports that were coming back from the conferences. These reports came from the IC designer to the poor, unsuspecting, IC user. It came in the form of a 1-bit digitizer that would output a 24-bit word. I once asked an audience of engineers in a technical seminar what 2^{24} was equal to. I expected one of the

The Basics Behind Analog-to-Digital Converters

geniuses in the crowd to quickly shout out "16,777,216". The actual answer that I got was "4^{12}". And now, the person who called me and asked me to recommend a 32-bit converter can actually see the light at the end of the tunnel.

Having a converter that is capable of converting to a resolution 16 million codes can be overwhelming at first, but let's back up and take the bird's eye view, then follow it with the details. With this approach, we can start with an intuitive level, which will get us a long way as we go into the forest to look for the trees.

Here Is How the Sigma-Delta ADC Works

The Sigma-delta ($\Sigma{-}\Delta$) converter is a 1-bit sampling system. A functional block diagram of an ADC $\Sigma{-}\Delta$ converter is shown in Figure 2.12. In this system, multiple bits are collected and then sent through a digital filter where there is a fair degree of mathematical manipulation performed.

An analog signal is applied to the input of the converter. This signal needs to be relatively slow because the $\Sigma{-}\Delta$ ADC samples the input signal multiple times; this technique is known as *oversampling*. The sampling rate is hundreds of times faster than the digital results at the output ports. Each individual sample is accumulated over time with the previous samples. This collection is averaged to achieve a statistical result of the sampled input signals.

The $\Sigma{-}\Delta$ ADC can be broken into four discrete segments (plus the serial interface). But keep in mind that you may have to modify this simple diagram for the individual converters that are using. For instance, your converter has been enhanced or simplified, but the basic operation of these converters is the same from device to device. A good, working block diagram is shown in Figure 2.12.

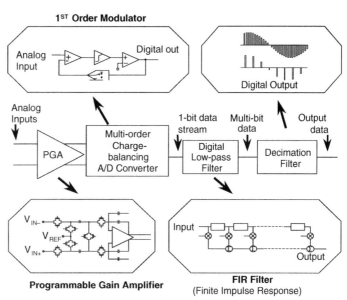

Figure 2.12: This block diagram of the sigma-delta converter has four segments, the programmable gain amplifier (PGA), multiorder charge-balancing ADC, a digital low-pass filter and a decimation filter.

Chapter 2

This figure has four basic blocks:

1. Programmable gain amplifier (PGA). Gain is achieved with capacitive double-sampling techniques (Figure 2.13).

2. Multiorder charge-balancing ADC. Charge is "balanced" across capacitors that surround an amplifier (Figure 2.14).

3. Digital low-pass filter which is often multiple finite impulse response (FIR) filters (Figure 2.15).

4. Decimation filter (Figure 2.16).

Programmable Gain Amplifier in the Σ–Δ Converter

The front-end of most Σ–Δ converters has a programmable gain amplifier (PGA). This is not the classical analog PGA implementation where there is an amplifier surrounded by resistors, which achieve an analog gain, purely in the analog domain. This PGA takes the analog input signal and quickly converts it to a sampled signal. The gain settings for this stage are programmed through the digital interface. In this stage, we haven't completely left the analog domain but we are halfway there. An example of this type of stage is shown in Figure 2.13.

The basic topology of the PGA stage is a differential switched capacitor amplifier. The switched capacitive topology uses a combination of oversampling and capacitor gain to achieve the possible gains of 1, 2, 4, 8 and 16. With this stage, the signal is clocked in on the rising edge of the sampling clock and transferred by the falling edge of the sampling clock to a second group of capacitors. For a PGA gain of one, the next rising clock edge sends the first signal forward to the modulator section of the A/D converter as well as sampling a second input signal. For PGA gains greater than one, say a gain of two, the second rising clock edge does not send the signal forward to the digital filter section. The second stage of the PGA retains the original signal and adds it to the second sampled input. In this manner the charge is doubled. At the completion of the second sampling, the charge is finally transferred to the modulator section. This concept can easily be extended to gains of 4, 8 and 16. As the gain of the PGA stage increases, the number of cycles required to sample the signal also increases. One of two scenarios can be implemented for this change in gain. The first scenario would keep a constant sampling clock and increase the overall sample time by 16× (for a gain increase of 16×). Another way that the PGA gain is implemented by increasing the number of samples taken by the input capacitor from 20 kHz for a gain of 1, to 320 kHz for a gain of 16. Adjusting the internal gain stage of the Σ–Δ converter is a technique that you can use to get an appropriate LSB voltage size for the transducer application.

Initially, switches G_1, G_2 and G_5 are closed, and G_3, G_4, G_6 and G_7 are opened on the rising edge of the clock. During the time that the clock is high, charge is accumulated on C_1 and C_2. On the falling edge of the clock, switches G_1, G_2 and G_5 are opened and G_3, G_4, G_6 and G_7 are

The Basics Behind Analog-to-Digital Converters

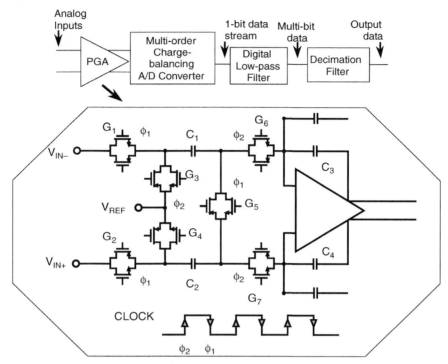

Figure 2.13: This switching network changes the analog input signal to a sampled voltage. This sampled voltage is taken in by the next stage where the signal is finally digitized into a one or zero.

closed. The charge on C_1 and C_2 is transferred to C_3 and C_4. This cycle is then repeated. The most common way to implement a gain is to sample the signal multiple times and store the charge on C_3 and C_4. For example, if the input signal charge is stored on C_3 and C_4 twice, the signal will have a gain of two applied.

Assuming the external clock is a constant 10 MHz, PGA gain is implemented by increasing the number of samples taken by the input capacitor from 20 kHz for a gain of 1, to 320 kHz for a gain of 16. Adjusting the internal gain stage of the Σ–Δ converter is a technique that you can use to get an appropriate LSB voltage size for the transducer application.

In digitizing systems, the anti-aliasing analog filter has saved many designs from noise-ridden disasters. They serve the purpose of rejecting high-frequency (uninvited) noise in the analog system so that the digitizer doesn't alias unwanted signals into the bandwidth of interest. One would assume that the anti-aliasing filter would always be a permanent fixture, placed before the analog-to-digital converter. With the sigma-delta converter, the internal digital network has nearly replaced this analog function.

Sigma-delta ADC manufacturers are promoting the requirement of a simple R/C low-pass filter at the input of the converter as the answer to all anti-aliasing problems. In fact, this

Chapter 2

filter does provide a small amount of high-frequency attenuation, but that is not the primary function of this simple low-pass filter. The most disruptive noise signals that are present at the input of the ADC are the switching currents coming in and out of the converter itself. The first stage of the unbuffered sigma-delta converter is fundamentally a switched capacitor network. A model of the Δ–Σ input is shown in Figure 2.14.

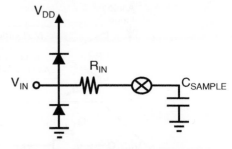

Figure 2.14: The input model of the sigma-delta converter is similar to the input model of the SAR converter. Consequently, if the external input impedance is too high, the offset and gain of a conversion can be compromised.

Switching glitches can be easily measured with a few hundred ohms on the inputs and an oscilloscope. This glitch energy can disrupt the measurement of the small voltages at the inputs by upsetting the driving input circuitry. This input stage should be treated with the same care that the SAR converter receives. Any error that is introduced because of high source impedance affects the offset and gain of the conversion. A step towards solving this problem is to place this R/C filter on the inputs of the converter. Additionally, if the device has differential inputs, a 0.1 µF capacitor can be placed directly across the inputs. This is done to attenuate high-frequency noise that is present at the input pins of the device. Note that this technique is not recommended for analog operational amplifiers.

The fact that this is called a PGA does not imply that input impedance if high. As a matter of fact, it is fairly low and dependent on the input capacitance and the over-sampling frequency.

Multiorder Charge Balancing ADC

The charge balancing ADC is the heart of this converter. It is responsible for digitizing the input signal and lowering the noise at lower frequencies. In this stage, the architecture implements a noise-shaping function where low-frequency noise is pushed up to higher frequencies. This low-frequency noise appears outside the band of interest. This is one of the reasons that Σ–Δ converters are well-suited for low-frequency, high-accuracy measurements.

The function of the multiorder charge-balancing A/D converter can be conceptualized with the 1st order stage shown in the insert in Figure 2.15. The analog input voltage and the output of the 1-bit DAC are differentiated, providing an analog voltage at X_2. The voltage at X_2 is presented to the integrator. The output of the integrator progresses in a negative or positive direction. The slope and direction of the signal at X_3 is dependent on the sign and magnitude of X_2. At the time the voltage at X_3 equals the comparator reference voltage, the output of the comparator switches from negative to positive or positive to negative, dependent on its original state. The output value of the comparator (X_4) is clocked back into the 1-bit DAC, as well as clocked out to the digital filter stage. At the time that the output of the comparator switches

The Basics Behind Analog-to-Digital Converters

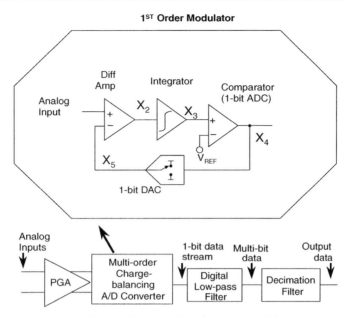

Figure 2.15: The multiorder charge-balancing stage of the Σ–Δ converter samples the input signal from the PGA stage and converts that signal to a one or a zero. Since this stage has a DAC in the feedback loop, the noise in this conversion is pushed out of the lower frequency band. This phenomenon provides the first step to an extremely low-noise conversion.

from a high to a low or vise versa, the 1-bit DAC responds on the next clock pulse by changing its analog output voltage to the difference amplifier. This creates a different output voltage at X_2, causing the integrator to progress in the opposite direction.

Many times the PGA stage of the multiorder charge-balancing stage is combined. But for discussion purposes, in Figure 2.15 the signal enters this stage from the PGA stage (Figure 2.13). Although this signal was sampled in the previous stage, the voltage magnitude can be anywhere between ground and the voltage reference. Depending on the 1-bit DAC's output voltage, the difference amplifier will produce a relatively high voltage or low voltage. This voltage is then integrated through the next stage (integrator). At the output of the integrator, a comparator will produce a one or zero. This is essentially the step where the signal is digitized, and in that step there is quantization noise created. The comparator output is sampled by the DAC in sync with the sampling clock of the converter. This may or may not change the DAC output. The output of the comparator is also sampled in sync with the sampling clock of the digital filter.

Chapter 2

The combination of the integrator and sampling strategy implements a noise-shaping filter on the digital output code. This noise shape is illustrated in Figure 2.16.

A digital filter is implemented in the next stage. The effects of the digital filter are illustrated in Figure 2.16.

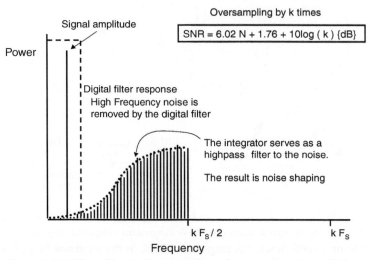

Figure 2.16: The noise in the modulator is moved out into higher frequencies. This is called noise shaping. After the modulator, the digital filter is used to implement a low-pass filter where high-frequency noise is removed. The calculation for the SNR includes the effects of oversampling.

Digital Low-Pass Filter

The output of the multiorder charge-balancing converter is a series of digital ones and zeros, which is sent to the digital filter. The digital filter uses an oversampling and averaging algorithm to further process the signal into the higher resolutions. The combination of the digital filter and the decimation filter stages directly affect the resolution and output data rate of the converter.

Typically, the digital filter is a finite impulse response (FIR) filter that essentially implements a weighted-average on the digital output from the modulator. A 1st order FIR filter is actually an averaging machine. A FIR digital filter is shown in Figure 2.17.

The 1st order FIR filter in Figure 2.17 uses a moving-average process. This averaging process mathematically reduces the level of uncertainty in the output signal, but it does take time to acquire the samples. Theoretically, if you acquire four samples, you can change the digital output from an output with two possibilities (0 and 1) to an output that has four possibilities (00, 01, 10, 11). This is possible through the oversampling mechanism and averaging process.

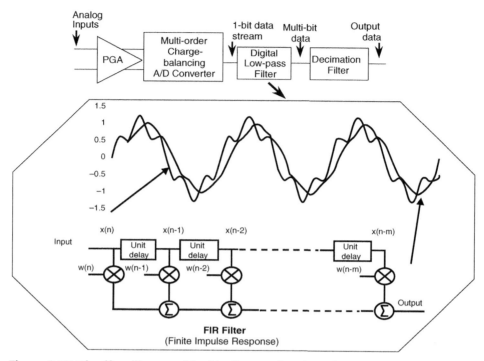

Figure 2.17: The filter illustrated in this diagram is a 1st order FIR filter. The digital low-pass filter of the Σ–Δ converter further reduces the noise in the conversion process by averaging multiple digital codes in this rolling-averaging filter (FIR filter).

For instance, if 4 bits are acquired, the possibilities, after the averaging process are 0, 0.25, 0.5 and 0.75. Each oversample by a factor of four gives a 6 dB (or 1-bit) improvement in the converters SNR. So theoretically, given a 1-bit ADC, a 2-bit converter can be mathematically realized with 4^1 averaged samples. A 3-bit converter can be realized with 4^2 (or 16) averaged samples. A 4-bit converter can be realized with 4^3 (or 64) averaged samples. You can see that this technique only works so far. For instance, if a 24-bit converter is derived from a 1-bit ADC, by averaging 4^{23} (or 70,368,744,178) samples. This would take a long time. As a result, there are other techniques used to obtain higher bit resolutions. These techniques include the modulator noise shaping (discussed previously), weighted averaging techniques, and multiple-level modulators.

There are very few Σ–Δ converters on the market that have a 1st order FIR filter. Most of these types of converters have a multiorder implementation of this function. The most common is 3rd order. You will also find that this 3rd order filter is called a $sinc^3$ filter. The transfer function of a $sinc^3$ filter with a cut-off frequency of 60 Hz is shown in Figure 2.18.

Chapter 2

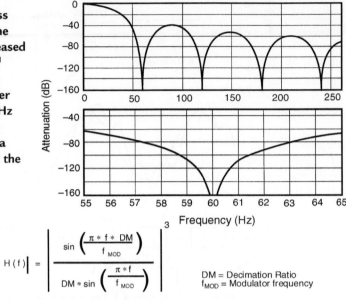

Figure 2.18: A low-pass FIR filter attenuates the noise slowly with increased frequency. This is a 3rd order FIR filter known as a sinc^3 low-pass filter and is tuned for a 60 Hz notch frequency. This type of filter provides a comb response across the frequency spectrum.

$$|H(f)| = \left| \frac{\sin\left(\frac{\pi * f * DM}{f_{MOD}}\right)}{DM * \sin\left(\frac{\pi * f}{f_{MOD}}\right)} \right|^3$$

DM = Decimation Ratio
f_{MOD} = Modulator frequency

Σ–Δ converters have a variety of digital filters to choose from. In this discussion, we have talked about the FIR filter. Another type of discrete-time filter that is common is the infinite impulse response (IIR) filter (Figure 2.19).

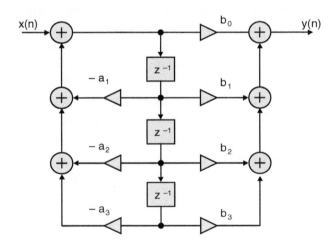

Figure 2.19: The infinite impulse response (IIR) digital filter has feedforward and feedback paths that assist in achieving sharper transitions regions, however, the trade-off for this advantage is possible instability.

Continuous-time filter approximations, such as Butterworth, Bessel, Inverse Chebyshev, and so forth, can be mapped into discrete-time IIR filters. But ultimately the design procedure for a discrete-time system begins with a set of discrete-time specifications. Continuous-time filter approximations are used as a convenient tool for meeting the discrete-time filter specifications. This type of discrete-time filter is transformed from the prototype continuous-time filter.

The advantage of the FIR filter is its stability and linear phase response. It has a simple, straightforward design (Figure 2.17) and uses relatively lower power than the IIR filter. On the down side, the FIR filter has to be a higher-order filter to get the job done, so the latency (the amount of time required for a good conversion) is longer. With the IIR low-pass filter, you can use lower order filters to accomplish the same cut-off frequencies as the FIR filter. As a result, there are few transistors in the implementation on silicon. On the down side, their filters have a nonlinear phase response and can be unstable.

Decimation Filter

The decimation filter is the final stage in the Σ–Δ converter, before the serial interface. The primary job of the decimation filter is to slow down the data output rate. This seems like it would not be a good idea, but in fact, the converter user is not interested in the intermediary averaging steps that the converter implements. Instead, the user is only interested in the final results.

So with this filter, the output is reduced considerably with respect to the sampling frequency.

The discrete, high-precision analog front-end in the data acquisition circuit is not out-dated, but is being gently pushed further into its exclusive corner. This is not to say that the demand for precision data acquisition circuits has disappeared. The solutions to these problems are

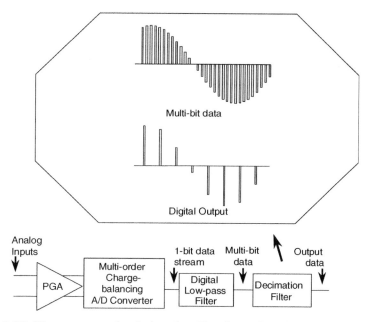

Figure 2.20: The purpose of a decimation filter is to slow down the output data rate in order to match the output data rate to the input frequency bandwidth. The FIR low-pass filter inputs a multibit data stream to the input of the decimation filter. The simplest decimation filter filters out several conversion results, while keeping enough to meet the output data rate of the converter.

changing in orientation from the analog-dominated circuit to digital. Although, the task of signal processing seems to be migrating to the digital domain, good analog engineering practices still apply. To my delight, it seems that the art of analog hardware design may be renamed but will never be obsolete.

What's the Big Deal About Σ–Δ?

The sigma-delta ADC is a new breed of device. Generally, this device samples analog signals at lower data rates than SAR converters, but it has the advantage of having higher resolution. With this higher resolution, the classical analog front-end can be eliminated. It is part of the initiative where digital designs are encroaching into the analog-domain hardware. Initially, this was done with firmware-controlled internal timers, comparators, and I/O gates, in conjunction with external resistors and capacitors. Some of the basic analog functions like D/A converters, ADCs and integrators have been realized. An example of a sigma-delta ADC design using a comparator, timer, external resistors and capacitors is discussed in Chapter 8. The implementations of these functions are primitive, but they get the job done when the accuracy and fidelity of high-precision analog functions are not needed.

The integrated sigma-delta ADC is quickly bridging this gap. Since the device is built using a complementary metal oxide semiconductor (CMOS), the available digital functions from controllers, processors and memory devices can be exploited. To enhance the features of this digital capability, more analog circuits are migrating from bipolar processes into the CMOS world. Not only is this migration occurring, but also the performance "quality" of these CMOS devices is improving. Some of the features that you can find in a sigma-delta ADC are shown in Figure 2.21.

The sigma-delta ADC integrates a significant quantity of analog and digital functionality. These two functions are designed to interact extremely well. For example, an analog buffer can be digitally switched in or out of the input stage. A multiplexer can be found on some sigma-delta ADCs, and the channels are programmed through the digital interface of the converter.

On the analog side of this device, voltage or current references are integrated into the chip, which allows for ratiometric operation. The advantage of a ratiometric system is that gain errors are eliminated because every element in the circuit uses the same reference. This applies to the excitation of the sensor all the way to the reference of the ADC. As mentioned before, there is usually a PGA on the front end of the converter, and some converters have buffers that can be switched in or out of the circuit digitally.

In terms of digital features, all converters have a serial interface where various modes can be programmed into the converter. For example, calibration algorithms can be implemented, the digital filter can be reset (or cleared) and status flags can inform the user about brownout events. Additionally, the digital filter corner frequency can be adjusted to match the application requirements.

The Basics Behind Analog-to-Digital Converters

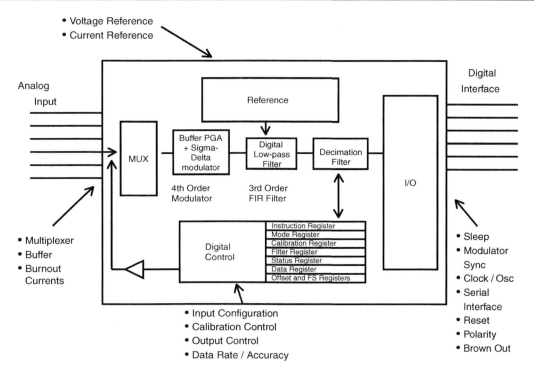

Figure 2.21: The sigma-delta ADC has matured to a point that there are analog functions that cater to a variety of sensor requirements as well as digital functions that integrate controller digital functions and features. This combination is only possible in CMOS. Some modules that can be added to the sigma-delta ADC are shown here. You won't find all of these features in any one sigma-delta ADC. This would consume too much silicon and consequently be too expensive where all of the features would not be used. Instead, you will find converters that use a select few of the features in order to target a converter to a specific application class.

Sigma-Delta ADC Specifications – Digital Filter Settling Time

One fundamental characteristic that separates this converter from the SAR converter is the settling time of the filter. Settling time with sigma-delta ADC defines the number of conversions that are required before the converter gives a reliable conversion. The settling time of the sigma-delta ADC is always a unitless integer and equal to the order of the digital filter. For example, the filter described earlier is a 3rd order FIR filter, otherwise known as a sinc^3 filter. The settling time of this filter is three. Quite simply, a 3rd order FIR filter has three 1st order, FIR filter stages with a serial configuration. This type of filter requires three conversions to completely push out the data contained in all three stages from the previous conversions. The result of this conversion process is shown in Figure 2.22.

As shown in Figure 2.22, it takes three complete conversions before the output code represents the input signal accurately. This dramatic example might be misleading in that it gives

Chapter 2

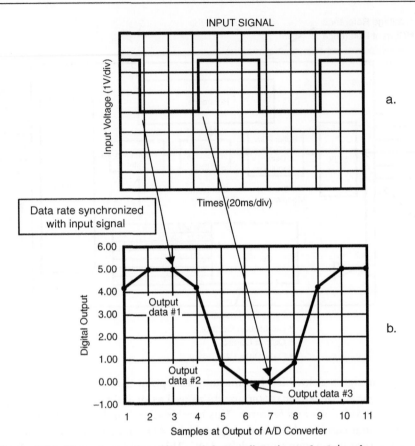

Figure 2.22: You can see the effects of the settling time of a 3rd order FIR filter if you synchronize an input square wave with the converter. The sampling starts for the first output data (b) on the falling edge of the input signal (a). The first output data code is approximately 80% too high. This result is a combination of one of the three filters containing the data from the actual input and the other two filters containing data taken with the previous input level of 5 V. The second output data code is now approximately 20% too high. This result is a combination of two of the three filters containing the data from the actual input and the other filter containing data taken when the input level was 5 V. The third output data code represents the input signal. All three stages of the digital filter contain data when the input signal is low.

the impression that every time you need an accurate conversion you will have to run three complete conversions. This is not exactly true in a real-life application setting. Usually the input signal going into a sigma-delta ADC is slow. Certainly much slower than a square wave. This is true unless you have a multiplexer on the input.

There is a class of converters that "hide" the conversions that stand a chance of being inaccurate. This is easily done by producing only output data #3 (in Figure 2.22). The consequences of this type of conversion strategy are that the overall data rate of this converter is slow, but you are always guaranteed an accurate result. This type of converter becomes very advantageous when you have an application where you want just one result. After you acquire that result, you shut the converter down.

Other Sigma-Delta ADC Differences from Vendor-to-Vendor

The sigma-delta ADC has opened the door for a large variety of features because of the digital side of the converter. In essence, the real estate that is dedicated to analog is about 25% of the silicon and digital occupies 75%. Having digital flexibility opens up Pandora's box. With the sigma-delta ADC, the feature set is limited because of silicon, not technology.

The fundamental features that most sigma-delta ADCs have are sleep states, slave/master settings, internal/external clock options, variable sampling frequency and voltage references. There are other features that some of the converters (but not all of them) have. For instance, some converters have an idle tone detect bit. Other converters have current references instead of the standard voltage reference; and yet other converters have self- and system-calibration capability, and the list goes on. Basically, the sigma-delta ADC is the first of many devices that will be integrated in analog and digital functions together under one roof.

Conclusion

In this chapter, we started by discussing the general key specifications for ADCs. We discussed how they impact the digital results from your converter. Then we looked at the successive approximation register (SAR) ADC and sigma-delta ADC. We took a strong look at the basic topologies of these converters and how they affect your conversion results. Now, going on to Chapter 3, we are going to apply these converters to "real-life" applications.

Chapter 2 References

Analog-Digital Conversion Handbook, Sheingold, Daniel H., Prentice-Hall, 1986.

Delta-Sigma Data Converters: Theroy, Design, and Simulation, Norsworthy, Schreier, Temes, IEEE Press, 1997.

"Voltage Reference Scaling Techniques Increase the Accuracy of the Converter as well as Resolution," Baker, Bonnie, Application Bulletin, AB-110, Burr-Brown Corporation, February, 1997.

1994 Application Seminar, Burr-Brown, Chapter 1.

"Understanding A/D Converter Performance Specifications," Bowling, Stephen, AN693, Microchip Technology.

"Using the Analog-to-Digital (A/D) Converter," Mitra, D'Sousa, Cooper, AN546, Microchip Technology.

"How to Get 23-Bits of Effective Resolution from Your 24-bit Converter," Baker, Bonnie C., AB-120. Burr-Brown Corporation, September 1997.

"Synchronization of External Analog Multiplexers with the $\Delta\Sigma$ A/D Converter," Baker, Bonnie C., Application Bulletin, AB-116, Burr-Brown Corporation, June 1997.

"Giving $\Delta\Sigma$ Converters a Little Gain Boost with a Front End Analog Gain Stage," Baker, Bonnie, Application Bulletin, AB-107, Burr-Brown Corporation, January, 1997.

"Switched-Capacitor A-D Converter Input Structures," Johnson, Jerome, AN30, Crystal.

"Delta-Sigma A-D Converter Conversion Technique Overview," AN10, Crystal.

"Using Sigma-Delta Converters, Part 1," AN388, Analog Devices.

"Using Sigma-Delta Converters, Part 2," AN389, Analog Devices.

"A Brief Introduction of Sigma-Delta Conversion," Jarmon, David, AN9054, Harris Semiconductor.

"Using Operational Amplifiers for Analog Gain in Embedded System Design," Baker, Bonnie C., AN682, Microchip Technology, Inc.

Discrete-time Signal Processing, Oppenheim, Schafer, Pentice-Hall, Inc., 1989.

"Number of Bits vs. LSB Errors," Baker, Bonnie C., EDN Magazine, July 8, 2004 (Reprinted with permission by EDN magazine, Reed Business Information, copyright 2004).

"Anticipate the Accuracy of Your Converter," Baker, Bonnie C., EDN Magazine, March 18, 2004 (Reprinted with permission by EDN magazine, Reed Business Information, copyright 2004).

CHAPTER 3

The Right ADC for the Right Application

CHAPTER 3

The Right ADC for the Right Application

Now that we have the ADC fundamentals out of the way, it's time to do some real work by looking at what these ADCs can do for us. In this chapter, we will spend some time looking at the places where signals come from, and which ADC is the best for the application. Through this exercise, we will have a better feel on how to prepare the signal for the microcontroller. Once this groundwork is established, we will move into looking at four specific applications. This chapter contains four real-world applications: temperature, pressures, light sensing and motor-control. In these applications, you will have to decide whether a SAR or Σ–Δ converter is appropriate. We will discuss the most common ADC problems and then quickly move to solutions. We will find that the SAR converter can service some of these applications. The sigma-delta ADC will better service others. Some applications can utilize both converters, with a minimum number of trade-offs.

Classes of Input Signals

Before we dive into the details of applications, we need to define the origin of real-world analog signals. From there, I will show you how to capture those signals with your circuits. Figure 3.1 shows several possible signal sources, with respect to the frequency versus number of bits.

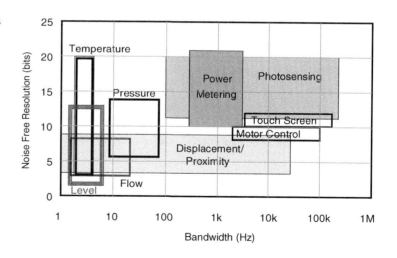

Figure 3.1: Input signals come to the ADC from various sources, but they mostly come from the physical world. The dynamics of these real-world signals define the sampling bandwidths. All of these physical sources have their own set of sensors, and these sensors define the number of required noise-free bits.

65

Chapter 3

At the system level, an appropriate ADC can produce the proper number of noise-free bits, or you can use an analog front-end gain cell plus a lower resolution ADC at a lower cost. This is a design architecture trade-off where cost, board space and number of components come into play.

When it comes to sensing external analog events, at the very least, you will want to measure temperature. Temperature testing is also the lowest common denominator in terms of frequency. This is because temperature doesn't change that quickly and the sensors reflect that characteristic. The physical environment of the sensor and the sensor package governs the speeds required for temperature-sensing events. Generally, the temperature changes in the environment are slow (> 0.1 sec/°C). This is consistent with the package and temperature-coefficient of the sensor. The sensor requires heat-up or cool-down periods if you want to make an accurate measurement.

Table 3.1: Temperature sensing applications use the thermocouple, RTD, thermistor or integrated circuit (IC) sensing elements. The inexpensive thermocouple does not require excitation, but requires a lookup table in the controller to linearize the results. The RTD sensor is the most accurate and operates over a wide temperature range, but requires current excitation and it may be cost-prohibitive for your application. The thermistor requires voltage excitation, with a ± Δ25°C linearity correction using series resistors. The silicon IC temperature sensor does not require external circuitry, but the accuracy is somewhat limited and the device is slow to respond to fast temperature changes.

	Thermocouple	RTD	Thermistor	Integrated Silicon
Temperature Range	–270 to 1800°C	–250 to 900°C	–100 to 450°C	–55 to 150°C
Sensitivity	10s of μV/°C	0.00385 Ω / Ω°C (Platinum)	Several Ω / Ω °C	Based on a technilogy that is ~2mV/°C sensitive
Accuracy	±0.5°C	±0.01°C	±0.1°C	±1°C
Linearity	Requires at least a 4th order polynomial or equivalent lookup table.	Requires at least a 2nd order polynomial or equivalent lookup table.	Requires at least 3rd order polynomial or equivalent look up table. Can also be linarize to 10-bit accuracy over a 50°C temperature range.	At best within ±1°C. No linearization required.
Ruggedness	The larger gage wires of the thermocouple make this sensor more rugged. Additionally, the insulation materials that are used enhance the thermocouple's sturdiness.	RTDs are susceptible to damage as a result of vibration. This is due to the fact that they typically have 26 to 30 AWG leads which are prone to breakage.	The thermistor element is housed in a variety of ways, however, the most stable, hermetic thermistors are enclosed in glass. Generally, thermistors are more difficult to handle, but not affected by shock or vibration.	As rugged as any IC housed in a plastic package such as dual-in-line or surface outline ICs.
Responsiveness in stirred oil	Less than 1 sec	1 to 10 secs	1 to 5 secs	4 to 60 secs
Excitation	None required	Current source	Voltage source	Typically supply voltage
Form of Output	Voltage	Resistance	Resistance	Voltage, current or digital
Typical Size	Bead diameter = 5 × wire diameter	0.25 × 0.25 in.	0.1 × 0.1 in.	From TO-18 Transistors to Plastic DIP
Price	$1 to $50	$25 to $1000	$2 to $10	$1 to $10

The Right ADC for the Right Application

The more common type of temperature sensors that you can use in your circuit is a thermocouple, resistance temperature devices (RTD), thermistors or integrated silicon sensors. Table 3.1 summarizes the general characteristics of the devices.

Some of you may wonder what these sensors look like. I find that a picture is worth a million words (see Figure 3.2). You will notice that size, and hence, the thermal bulk is quite big. Again, this is only because temperature typically changes very slowly.

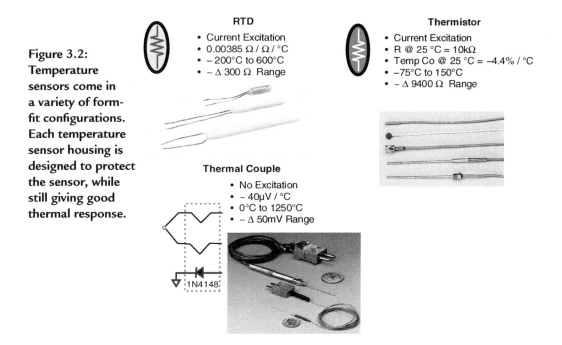

Figure 3.2: Temperature sensors come in a variety of form-fit configurations. Each temperature sensor housing is designed to protect the sensor, while still giving good thermal response.

RTD
- Current Excitation
- 0.00385 Ω / Ω / °C
- −200°C to 600°C
- ~ Δ 300 Ω Range

Thermistor
- Current Excitation
- R @ 25 °C = 10kΩ
- Temp Co @ 25 °C = −4.4% / °C
- −75°C to 150°C
- ~ Δ 9400 Ω Range

Thermal Couple
- No Excitation
- ~ 40µV / °C
- 0°C to 1250°C
- ~ Δ 50mV Range

The granularity of the temperature measurements in terms of bits-per-degree Celsius can be large or quite small, as shown in Table 3.1. Additionally, the number of bits that are required for your system can be low or relatively high. Of course, this depends on your requirements. Because there is a wide range of application conditions, you can use the SAR converter or the Σ–Δ converter for temperature measurements. There are many ways to implement the signal conditioning circuitry of a temperature sensor. Later in this chapter, we will examine SAR and Σ–Δ ADC options with an RTD sensor.

The other physical entities in Figure 3.1 are level, flow and displacement or proximity. You usually don't measure flow directly with a sensor that specifically measures flow. You can measure flow with temperature sensors or by measuring vibrations in a tube filled with a fluid. Displacement and proximity are other entities that don't have specific sensors dedicated for that purpose. You can measure these two phenomena optically with an LED and photodetector. You can also use an accelerometer.

Chapter 3

The sensors for motor control, touch screen and power metering are typically resistive. For example, you can place a very small resistor (tenths of ohms) in the "legs" of the motor control, MOSFET switching lines. The sensors of touch screens are generally resistive or capacitive. The resistive touch screen has gained attention because of the personal data assistants (PDA), and capacitive screens are used in "dirty' environments. In all cases, an ADC eventually digitizes the signal from the sensor. You can accomplish this with a SAR or sigma-delta ADC.

The characterization (Figure 3.1) of different measurements in terms of sampling frequency versus number of bits is instructive to a point. One question that remains in this discussion is, "What are the output voltage ranges of these sensors?" Figure 3.3 answers this question.

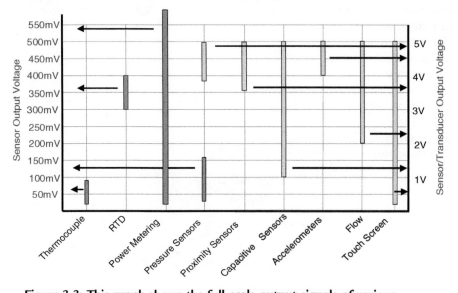

Figure 3.3: This graph shows the full-scale output signals of various sensors. It uses, where appropriate, a 1 mA or 5 V excitation. The output voltage level of various sensors, along with the sampling frequencies summarized in Figure 3.1 will play a large part in determining which ADC you choose for the application circuit. There are two classes illustrated in this figure. The first class produces millivolt outputs across the sensors full-range. The left y-axis refers to this group of sensors. The second class produces voltage outputs across the sensors full range. The right y-axis refers to the output ranges of this group. Most of the second group of devices are transducers.

There are two classes of output voltage ranges for sensors. The x-axis lists all of the sensors. In the first class, the output voltages of the sensors range from approximately 50 mV to 600 mV. The scale for this range is on the left y-axis in Figure 3.3. The right-most y-axis sensors are the thermocouple, resistance temperature device (RTD), power metering and pressure sensors.

The second class of sensor-outputs range from about 0.25 V to 5 V. The y-axis for this output range is on the right side of the graph as shown in Figure 3.3. Generally, the sensors that fit into this output range are pressure sensors, capacitive sensors, accelerometers, flow meters and touch screens. You probably will note that pressure sensors fall into both classes of voltage output ranges. The pressure sensor in the first class is strictly a sensor, and the pressure sensor in the second class is a pressure sensor with an integrated signal conditioning system on-chip.

As a foundation, you should have a grasp of ADC specifications. You should also know which specifications are the most important for your application. This knowledge can save an enormous amount of time during your design phase by simply paying attention to only the important details, not the whole list of performance characteristics and specifications.

The next section in this chapter gives you the ammunition to gain a basic understanding of how the SAR converter and Σ–Δ converter work in your sensor circuits. With this knowledge, you can quickly determine which ADC is right for your application. For example, if your ADC reports data at the wrong time, too fast, too slow or too inaccurate, the controller or processor will struggle with too much data, not enough information or erroneous data. The ADC can be the genius behind the system or its downfall.

Temperature Sensor Signal Chains

The SAR converter and Σ–Δ converter are similar enough in performance in that they can both be applied to the same temperature circuits. Regardless of the converter you have chosen for your sensor, your circuit will always require some degree of analog circuitry. However, these two converters are different enough so that one or the other, not both converters, can only serve some sensor applications. Of the two devices, the SAR converter is the easiest to use and understand, but the Σ–Δ converter has more features and functions. The Σ–Δ converter has enough features to make you a hero or get you into serious trouble (as discussed in Chapter 2). Figure 3.4 illustrates two system configurations where SAR converters and/or Σ–Δ converters are used.

In Figure 3.4a and b, the input sensor is a resistive Wheatstone bridge. I used the Wheatstone bridge out of convenience in this diagram. Other sensors can take the place of the Wheatstone bridge in these circuits. For example, an array of temperature sensors, optical sensors and capacitive sensors (to name a few) would fit the bill. This resistive bridge in Figure 3.4 can model a variety of sensors, most commonly the pressure sensor or load cell.

In the top signal chain (Figure 3.4a), the AMP block differentiates and amplifies the two analog, output signals from the bridge. This AMP block can just be operational amplifiers or an instrumentation amplifier, depending on the type of sensor in the circuit. In this system (Figure 3.4), the instrumentation amplifier will simultaneously sample the signals from the bridge and eliminate most of the common-mode noise. (Refer to Chapter 6 for instrumentation amplifier discussions.) Many times, there is a voltage reference (V_{REF}) attached to the

Chapter 3

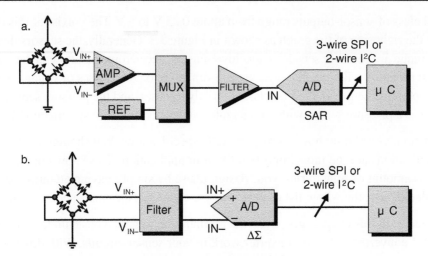

Figure 3.4: With many applications, you can use the SAR ADC or the sigma-delta ADC to digitize the analog signal. Many times the SAR ADC requires front-end analog gain circuitry. The sigma-delta ADC usually does not require this additional circuitry, but its slower speed will sometimes eliminate this type of device as an option.

amplifiers or instrumentation amplifier. These voltage references come in handy in single-supply circuits. They can provide a reference point to the center of your supplies. They also are helpful when you want to achieve a ratiometric relationship between the signal and the ADC conversion.

Next in the signal line, in Figure 3.4a, is a MUX block, or multiplexer. It is not unreasonable in this type of system to have several signals digitized by the same ADC. You will frequently find a need to measure temperature, which will consume one channel of the multiplexer. Sensing other physical entities like pressure, force, light and so on can use any one of the other channels.

You may or may not use a multiplexer in this signal path, but you will need an analog filter. You can implement the FILTER block one of three ways (in Figure 3.5). There are more details about filters in Chapter 4, but I want to make a few sweeping statements at this point. The first and most fundamental way to design an analog filter is to use resistors and capacitors (Figure 3.5a). As we will see in more detail in Chapter 4, this filter can compromise the signal because of impedance matching problems. As a second alternative, you can design an active low-pass filter by using an operational amplifier, a few resistors and capacitors (Figure 3.5b). A third, and maybe more attractive solution for the digital designer is to use a switched-capacitor filter.

Although we will cover these filters in detail in Chapter 4, I want to emphasize that a switched-capacitor filter is not your usual analog filter. They are sampling devices that keep

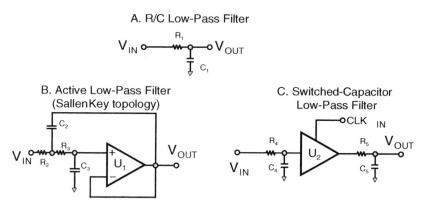

Figure 3.5: An easy low-pass filter design uses resistors and capacitors (a). The pole created by this filter is equal to $1/(2\pi RC)$. Multi-pole filters can use an active-filter topology (b). This 2nd order filter can be cascaded with more filter stages to build a higher-order filter. Switched-capacitor filters (c) are multi-pole, low-pass filters with very few external components. This type of low-pass filter can be a 5th (or higher) order filter.

the signal in the analog domain, but also have the same aliasing problems as the ADCs. If you don't use special care with switched-capacitor filters, you may introduce as much or more noise than you are trying to eliminate. The primary function of the analog filter is to eliminate the higher frequency signals that can be aliased by the ADC. You will usually need a 4th, 5th or 6th order low-pass filter in this signal path.

The analog FILTER precedes the ADC. The ADC digitizes the signal, which is free of higher frequencies because of the filter. Since there is a pre-gain stage in this signal path (AMP block), the ADC will most likely be a SAR converter. However, sometimes you might use a $\Sigma{-}\Delta$ converter because the input-signal voltage is so low. With the SAR converter, the ADC samples the signal once per conversion before digitization.

Following the ADC is a microcontroller (µC block). This block could also be a processor or field programmable gate array (FPGA). Many times the controller or processor will have an internal SAR converter. If you have a controller that has this function integrated, it can replace your stand-alone converter as long as you understand the possible errors introduced. An integrated ADC in the processor or controller runs the risk of being noisy because of the digital switching on the chip. Another issue with internal ADCs is that the IC manufacturers can have silicon limitations while they are trying to control costs. If this is the case, you can compromise the accuracy of various cells on the integrated chip.

The second signal path (Figure 3.4b) looks simpler than Figure 3.4a. In this signal path the sensor signal goes through a filter block, into an ADC and then to the controller. Typically, the FILTER block is less complex than the FILTER block in Figure 3.4a. It is usually a signal-pole filter built with an R/C pair. Once the signal is through the filter, it goes to the ADC. Note that the input of the ADC is differential. The input differential stage of the ADC performs that same task as the analog instrumentation amplifier, which it rejects common-mode noise.

Since this signal path seems to be simplified, you might think that it is your best option. As it turns out, in this signal path, the ADC is usually a Σ–Δ converter. Your circuit layout complexity increases if you add the Σ–Δ device. Usually multiple layer boards are needed with sigma-delta ADCs where the SAR converter could easily reside on a two layer board (layout details are discussed in Chapter 11). The various blocks in Figure 3.4a have been absorbed into the Σ–Δ converter. For instance, many times Σ–Δ have a multichannel, multiplexer input stage.

Using an RTD for Temperature Sensing: SAR Converter or Sigma-Delta Solution?

Before you pick the interface circuit for the temperature sensor, you need to know the sensor basics. The platinum RTD temperature-sensing element is the most accurate temperature sensor available. It is also more stable over time and temperature than the other types of temperature sensors. RTD element technologies are constantly improving, which further enhance the quality of this temperature measurement. Typically, a data-acquisition-system conditions the analog signal from the RTD sensor, making the analog translation of the temperature usable in the digital domain as shown in Figures 3.7 and 3.8.

The acronym "RTD" means "resistance temperature detector." The most stable, linear, and repeatable RTD is made of platinum metal. The temperature-coefficient of the RTD element is positive. This is in contrast to the NTC Thermistor that has a negative temperature-coefficient. An approximation of the platinum RTD (PRTD) resistance-changes over temperature can be calculated by using the constant 0.00385 Ω/Ω/°C. This constant is easily used to calculate the absolute resistance of the RTD at temperature with the formula below:

$RTD(T) = RTD_0 + T * RTD_0 * 0.00385$ Ω/Ω/°C

where $RTD(T)$ is the resistance value of the RTD element at temperature in Celsius,

RTD_0 is the specified resistance of the RTD element at 0°C, and

T is the temperature of the environment where the RTD is placed

Typical specified 0°C RTD values are 50, 100, 200, 500, 1000 or 2000 Ω. Of these options, the 100 Ω platinum RTD is the most stable over time and linear over temperature.

Since resistors are hard to measure directly, the RTD element requires excitation. A constant current source (as opposed to a voltage-source) will provide a linear resistance to voltage conversion. When you excite the RTD element with a current source, the accuracy can be as good as ± 4.3°C over a –200°C to 800°C temperature range. If the magnitude of the current source is too high, the element will self-heat, which will cause temperature measurement errors. Therefore, care should be taken to ensure that ≤ 1 mA of current is used to excite the element.

The Right ADC for the Right Application

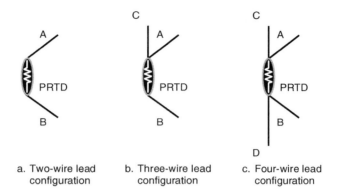

a. Two-wire lead configuration
b. Three-wire lead configuration
c. Four-wire lead configuration

Figure 3.6: This figure shows the three lead configurations for RTD elements. Figure a.) Lead configuration provides one connection of each end of the sensor. This configuration is prone to absolute and temperature errors due to the lead resistance. The three-wire (b.) configuration is the most commonly used configuration. This extra lead from the RTD element eliminates errors caused by currents through the leads. The most robust lead configuration is the four-wire lead (c.). Wires A and B connect to the high impedance input of the front-end circuitry and wires C and D conduct the excitation current.

The RTD Current Excitation Circuit for the SAR Circuit

For best linearity, the RTD sensing element requires a stable current reference. Figure 3.7 illustrates one way to implement a stable current reference. In this circuit, a voltage reference, along with two operational amplifiers generates a floating, 1 mA current source.

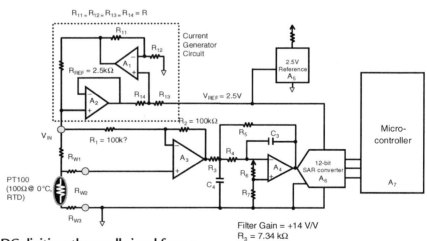

Figure 3.7: A SAR ADC digitizes the small signal from a three-terminal, RTD temperature sensor as long as an amplifier circuit gains this small signal from the RTD. In this circuit, the operational amplifier, A_4, is used to implement a gain as well as a 2nd order low-pass filter.

Filter Gain = +14 V/V
R_3 = 7.34 kΩ
R_4 = 22.4 kΩ
R_5 = 102 kΩ
R_6 = 1 kΩ
R_7 = 9 kΩ
C_3 = 0.033 µF
C_4 = 47 µF

The evaluation of this circuit starts with the 2.5 V precision voltage reference (A_5, upper right corner of Figure 3.7). A_5 connects directly to R_{13}. Since the noninverting input of the CMOS amplifier, A_1, is high impedance, the voltage drop across R_{13} and R_{14} is equal. The voltage at inverting input of A_1 is equal to the noninverting input voltage. That voltage is gained by $(1 + R_{11}/R_{12})$ to the output of the amplifier and the top of the reference resistor, R_{REF}. The voltage at the output of A_1 is equal to:

$$V_{OUTA1} = (1 + R_{11}/R_{12}) * (V_{REF} - V_R)$$
$$= 2 * (V_{REF} - V_R)$$

where V_{OUTA1} is the voltage at the output of A_1, and
V_R is the voltage drop across R_{13}.

If $R_{11} = R_{12}$, the voltage at the output of A_2 is equal to $V_{REF} - 2V_R$. This same voltage appears at the inverting input of A_2 and across to the noninverting input of A_2.

Solving these equations, the voltage drop across the reference resistor, R_{REF} is equal to:

$$V_{RREF} = V_{OUTA1} - V_{OUTA2}$$
$$= 2 * (V_{REF} - V_R) - (V_{REF} - 2V_R)$$
$$= V_{REF}$$

where V_{RREF} is the voltage across the reference resistor, R_{REF}, and
V_R is the voltage drop across R_3 and R_4.

The current through R_{REF} is equal to:

$$I_{RTD} = V_{RREF} / R_{REF}$$

This circuit generates a DC, floating, current source that is ratiometric to the voltage reference and the ADC.

Absolute errors in the circuit will occur as a consequence of the absolute voltage of the reference, the initial offset voltages of the operational amplifiers, the output swing of A_1, mismatches between the resistors (R_{11} R_{12}, R_{13} and R_{14}), the absolute resistance value of R_{REF}, and the RTD element. Errors due to temperature changes in the circuit will occur as a consequence of the temperature drift of the same elements listed above. The primary error sources over temperature are the voltage reference (A_5), offset drift of the operational amplifiers (A_1 and A_2), and the RTD element.

RTD Signal Conditioning Path Using the SAR ADC

You can digitize changes in resistance of the RTD element over temperature with a SAR ADC. Figure 3.7 shows a current-excitation circuit for an RTD element. With this style of excitation, you can tune the magnitude of the current source to 1 mA or less by adjusting R_{REF}. With this circuit, a three-wire RTD element is selected (see Figure 3.7). This configuration minimizes errors due to wire resistance and wire resistance drift over temperature. A_3 senses the voltage drop across the RTD element. A_4 then gains and filters the analog signal in preparation for the ADC input.

In the circuit using a SAR converter (Figure 3.7), the RTD element equals 100 Ω at 0°C. If the RTD senses temperature over from −200 to 600°C, the range of resistance from the RTD is nominally 23 Ω to 331 Ω. Since the RTD resistance range is relatively low, wire-resistance and wire-resistance change-over-temperature can skew the measurement of the RTD element. The three-wire RTD device reduces these errors.

The operational amplifier that contains A_3 subtracts wire-resistance error of R_{W1} and R_{W3}. In this configuration, R_1 and R_2 are equal and relatively high. You should select the value of R_1 to ensure that the leakage currents through the resistors do not introduce errors to the RTD element. The transfer function of this portion of the circuit is:

$$V_{OUT:A3} = (V_{IN} - V_{W1})(1 + R_2 / R_1) - V_{IN}(R_2 / R_1)$$

where $V_{IN} = V_{W1} + V_{RTD} + V_{W3}$,
V_{Wx} is the voltage drop across the wire to and from the RTD, and
$V_{OUT:A3}$ is the voltage at the output of A_3.

If $R_1 = R_2$ and $R_{W1} = R_{W3}$, the previous equation reduces to:

$$V_{OUT:A3} = V_{RTD}$$

A 2nd order, low-pass filter removes higher frequency noise from the voltage signal at the output of A_3. This filter is built using A_4, R_3, C_3, R_4, R_5 and C_4. The low-pass filter in this circuit should have a cut-off frequency that is as low as possible. This reduces amplifier and conducted noise as much as possible. You should choose the Chebyshev filter (0.5 dB ripple) because of its fast transition region in the frequency domain (see Chapter 4, "Do I Filter Now or Never?"). The closest expected high frequency noise in this circuit is 60 Hz. The attenuation of this filter at 60 Hz is 39 dB down from DC. The noise from the CMOS amplifiers could be as high as 29 nV√Hz (rms) @ 10 kHz. The noise due to the two amplifiers in the circuit signal path will be 48 µV (rms) or 0.318 mV (p-p) at the input to the 12-bit A/D converter. With a 10 Hz, 2nd order Chebyshev filter, only ~2 µV (p-p) remain in the signal. More critically, the noise that is injected by the mains frequency (50 Hz or 60 Hz) is reduced by –24.5 dB or –27.9 dB (inclusive). This is equal to an attenuation of 16.80× or 23× inclusive (see Chapter 10, "Noise – The Three Categories: Device, Conducted and Emitted").

The LSB size of the 12-bit converter equal to:

$$\begin{aligned} ADC_{LSB} &= V_{REF}/2^{12} \\ &= 2.5\ V/4096 \\ &= 0.610\ mV \end{aligned}$$

The sample speed of the SAR ADC can be as slow or as fast as need be. Your selection of this sampling speed will not affect the accuracy of the conversion.

Is the SAR ADC Right for this Temperature Sensing Application?

The SAR ADC is a good fit for this type of application. You need to be willing to carefully gain the voltage from the resistive RTD, but the gain cells are easy to implement. The biggest challenge in this circuit is the current reference circuit. The SAR converter offers a low power, low cost solution that is easy to implement into this circuit.

RTD Signal Conditioning Path Using the Sigma-Delta ADC

Sigma-delta A/D converters have an innate ability to resolve an analog input signal to a very small LSB voltage size. At first glance, high resolution doesn't seem to be an important specification for the RTD temperature sensor as long as you use an analog gain stage. To the contrary, close inspection of the interface circuit divulges a different story. The sensing element's output could be in the hundreds of millivolts. Worse yet, the output voltages that represent a change in temperature can be extremely low (sub-mV or µV). If the dynamic resolution of the A/D converter alone is relatively high, the total device count is lower by removing the front-end gain stage and reducing the complexity of the anti-aliasing filter (Figure 3.8).

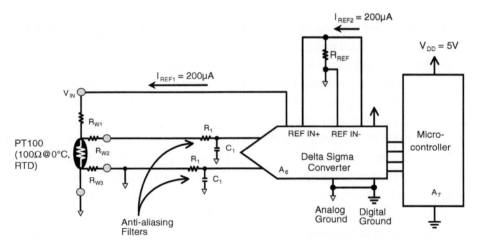

Figure 3.8: A Σ–Δ ADC directly digitizes the small signal from a four-terminal, RTD temperature sensor. In this circuit, a 200 µA current reference from the Σ–Δ ADC excites the temperature sensor. The 200 µA current source is ratiometric to the ADC voltage reference. The resistor/capacitive pairs (R1/C1) are used as anti-aliasing, low-pass filters.

For the given RTD of 100 Ω (@ 0°C) with an excitation current of 200 µA, the nominal full-scale output voltage range (−200°C to 600°C) would be 66.2 mV. The digitizing system that follows the RTD should reliably represent the temperature to 12-bit accuracy (given the error contributions of the RTD). The designer can choose to gain and actively filter the RTD voltage using analog techniques or put a digital engine to work; the sigma-delta converter. This predominantly digital solution uses a passive anti-aliasing filter followed by the Σ–Δ device that gains and filters the signal with digital computational methods. The sigma-delta converter makes this precision application possible because of its superior digital processing and system calibration capability.

The circuit in Figure 3.8 combines a high precision, four-wire lead PRTD with a sigma-delta converter. The reference current from the Σ–Δ converter excites the RTD, using two of the four pins. Using the remaining two wires of the thermal element, the voltage signal across the RTD is sensed using the differential inputs of the Σ–Δ ADC. This technique establishes a ratiometric relationship between the voltage reference to the ADC and the RTD temperature sensor.

In this application, I am using the AD7713 Σ–Δ ADC from Analog Devices, Inc. This device is perfectly suited for RTD sensing circuit, because of the two current references. Additionally, this device will generate a 24-bit code for low-level, voltage signals. This Σ–Δ ADC has self-calibration, system calibration, and background calibration options.

The internal Σ–Δ noise reduction techniques improve the performance of the circuit. The data rate of the converter is 10 Hz (as programmed) to reduce interference from the mains frequency, as well as reduce aliasing of higher frequency noise into the final signal output. This low frequency data rate is possible because of the slow responsiveness of the RTD element to temperature.

If the layout is good, the sigma-delta ADC will have an effective resolution of 20 bits (rms) or 17.27 noise-free bits. If that is the case, you can select a subset of the possible bits. For instance, a 17-bit conversion has 32 different places where 12-bits can be extracted. This is how you eliminate the gain stage in Figure 3.2 of the SAR converter circuit.

Is the Sigma-Delta ADC Right for this Temperature Sensing Application?

The answer is yes. The sample rate of this type of converter matches the sensor and the physical event. Since this converter will give you a high number of noise-free bits, you can ignore the ones that don't fit into your range of interest and still get a 12-bit conversion out of the deal.

This is an example of how the sigma-delta ADC is best suited for an RTD temperature measurement application. Although the sigma-delta ADC is appropriate here, a different application with a different sensor may require a different Σ–Δ converter or a SAR ADC.

Measuring Pressure: SAR Converter or Sigma-Delta Solution?

The second most common physical entity that we tend to want to measure is pressure. Pressure measurement devices can be classified into two groups: those where pressure is the only source of power, and those that require electrical excitation. The mechanical style devices that are only excited by pressure, such as bellows, diaphragms, bourdons, tubes, or manometers, are usually suitable for purely mechanical systems. With these devices, a change in pressure

Chapter 3

will initiate a mechanical reaction; for example, a change in the position of mechanical arm or the level of liquid in a tube.

Electrically excited pressure sensors are most suited to a microcontroller or microprocessor environment. These kinds of sensors can be piezoresistive, linear variable differential transformers (LVDT), or capacitive sensors. Usually, when you measure pressure you use the piezoresistive sensor.

The Piezoresistive Pressure Sensor

Pressure changes are usually slow, so lower speed conversions are acceptable. Another characteristic of measuring pressure is that the actual sensor itself has a differential output device. This differential output is very helpful when you want to reject common-mode environmental noise. Therefore, differential analog input devices are very useful.

The piezoresistive is a solid-state, monolithic sensor that has silicon processing. Piezo means pressure; resistance means opposition to a DC current flow. There are 300 to 500 piezoresistive pressure sensors per wafer. Since these wafers generate a large number of sensors, they are less expensive than mechanical sensors.

A pressure reference consists of a cavity that is fabricated from two wafers sealed together. The topside of this fabricated sensor is the resistive material, and the bottom is the diaphragm. The high side of the piezoresistive bridges shown in Figure 3.9 can have a voltage excitation or current excitation applied. Although the magnitude of excitation (whether it is voltage or current) affects the dynamic range of the output of the sensor, the maximum difference between V_{OUT+} and V_{OUT-} generally ranges from tens of millivolts to several hundred millivolts. The electronics that follow the sensor change the differential-output, sensor signal to a single-ended signal. The signal chain then gains and filters the signal in preparation for digitization.

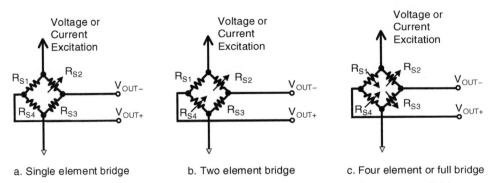

Figure 3.9: The resistive Wheatstone bridge configuration can have one variable element (a), two elements that vary with excitation (b), or four elements (c).

The model for this sensor is a four-resistor element, the Wheatstone bridge (Figure 3.9). Four-element bridges have better sensitivity as compared to a single element or two element sensors. By applying a positive differential pressure to the four-element bridge, two of the elements respond by compressing and the other two change to a tension state. When you apply a negative differential pressure to the sensor, the diaphragm moves in the opposite direction. Additionally, the resistors that were compressed go into a tension state, while the resistors that were in a tension state change into a compression state. Piezoresistive pressure sensors may or may not have an internal pressure reference.

The Pressure Sensor Signal Conditioning Path Using a SAR ADC

Figure 3.10 shows an example circuit of how to interface with this type of sensor with a SAR ADC.

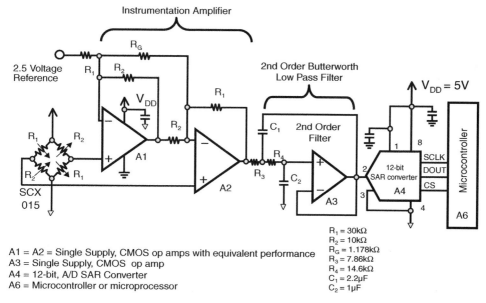

Figure 3.10: The voltage at the output of the SCX015 pressure sensor is gained (Gain = 37.7 V/V) by the instrumentation amplifier (A1 and A2). Then, a 2nd order 10 Hz, low-pass filter (A3) filters the signal. Finally, a 12-bit ADC (A4) digitizes the signal.

There are several ways of capturing the small, differential, output-signal of the sensor and transforming it into a usable digital code. One approach is to take the small differential output of the bridge, gain it, and convert it from differential to single-ended with an instrumentation amplifier (Figure 3.10). The signal then passes through a 2nd order, 10 Hz, low-pass filter. The low-pass filter eliminates out-of-band noise and unwanted frequencies in the system before the A/D conversion occurs. The stand-alone, 12-bit ADC follows the low-pass filter. The ADC transforms the analog signal into a usable digital code. The microcontroller takes the converter code, further calibrates, and translates if need be for display purposes. In this signal path, one analog filter is required. This analog filter follows the instrumentation amplifier.

Chapter 3

Is the SAR ADC Right for this Pressure Sensing Application?

This circuit configuration requires a fair amount of analog circuitry, prior to the ADC (Figure 3.10). However, if the application requires a higher sample rate, this converter may give you the best solution.

Pressure Sensor Signal Conditioning Path Using a Sigma-Delta ADC

A second way that the differential signal from the SCX015 pressure sensor is captured uses a $\Sigma-\Delta$ ADC as the core converter in the circuit. In this circuit (Figure 3.11), a single-pole, low-pass filter removes high-frequency noise for the differential output of the pressure sensor (SCX015). Immediately following these filters, the $\Sigma-\Delta$ ADC accepts the differential signal at its input. The resolution of the $\Sigma-\Delta$ ADC is considerably higher (usually 24-bits) than the resolution of the SAR converter in Figure 3.10. You can use this converter in the same manner as the converter in Figure 3.8. If you will remember, we did not use all of the bits at the output of the $\Sigma-\Delta$ in Figure 3.8. We only selected the 12-bit that we needed.

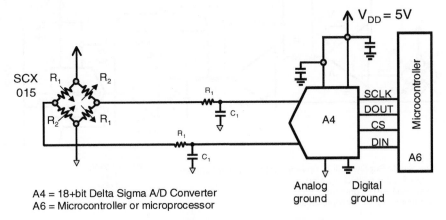

Figure 3.11: The voltage at the output of the SCX015 pressure sensor is filtered at both inputs to the sigma-delta ADC by single-pole low-pass filters (R1/C1) and digitized by the sigma-delta ADC (A4).

Is the Sigma-Delta ADC Right for this Pressure Sensing Application?

Bridge sensors are a perfect fit to the $\Sigma-\Delta$ ADC (Figure 3.11). The converter is able to accept a differential input while rejecting common-mode noise. This alone replaces one of the major analog functions of the instrumentation amplifier illustrated in Figure 3.10. Additionally, the number of bits of this type of converter allows the designer to eliminate the analog gain stage making this type of device a perfect fit.

Photodiode Applications

A close third on the popularity list of sensing circuits is the photosensing application. Photodiodes bridge the gap between light and electronics. Many times precision applications such as CT scanners, blood analyzers, smoke detectors, position sensors, IR pyrometers and chromatographs utilize the basic transimpedance amplifier circuit that transforms light energy into a usable electrical voltage. These circuits use photodiodes to capture the light energy and transform it to a small current. This current is proportional to the level of illumination from the light source. A preamplifier then converts the current (in amperes) from the photodiode sensor into a usable voltage level.

Photosensing Signal Conditioning Path Using a SAR ADC

A practical way to design a precision photosensing circuit is to place the photodiode in a photovoltaic mode. A photodiode, in its photovoltaic mode, has zero volts across it. This can be done by placing the device across the inputs of an amplifier and a resistor/capacitor pair in the feedback loop. Figure 3.12 shows a single-supply circuit implementation of a photosensing circuit.

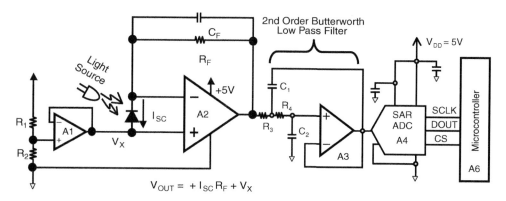

A1 = A2 = single supply CMOS amplifier with similar frequency performance
A3 = single supply CMOS amplifier
A4 = 12-bit SAR Analog-to-Digital Converter
A6 = Microcontroller or microprocessor

Figure 3.12: The light source impinging on a photodiode generates a reverse current from the diode through the feedback resistor, Rf. The voltage diode's anode as well as the non-inverting input of the amplifier are raised at a higher voltage than ground (~300 mV) to avoid amplifier output clamping near ground. A low-pass filter removes higher frequencies embedded in this output signal of the amplifier. The SAR ADC then receives this signal.

In this SAR photodetector circuit (Figure 3.12), the light source illuminates the photodiode, causing diode current to flow from cathode to anode. Since the input impedance of the inverting-input of the CMOS amplifier is extremely high, the current generated by the photodiode flows through the feedback resistor, R_F. The current to voltage transfer function of this circuit is:

$$V_{OUT} = I_{SC} * R_F + Vx$$
with a single-pole at $1/(2\pi R_F C_F)$

where:
V_{OUT} is the voltage at the output of the operational amplifier in volts.
I_{SC} is the current produced by the photodiode with units in amperes.
R_F is the feedback resistor with units in ohms.
C_F is the feedback capacitor with units in farads.

Once the signal becomes a voltage, it is easy to get a digital representation with an ADC. SAR converters are well-suited for this circuit because of their higher sampling speed. There are some current-input, ADCs on the market that also support this type of application.

In this circuit, the amplifiers A1 and A2 should have similar AC performance and noise specifications. It is best to make them the same amplifier. I am choosing the capacitor value, C_F, so that the amplifier circuit (A2) is stable. If you need help calculating the capacitor for this circuit, refer to Chapter 6, "Putting the Amp into a Linear System."

Is the SAR ADC Right for this Photodetection Application?

The SAR ADC is very appropriate for this application because of its conversion speed, compared to the sigma-delta ADC. The conversion speeds of the SAR converter report the optical events more reliably, where the $\Sigma\text{-}\Delta$ ADC will generally report DC type signals.

Photosensing Signal Conditioning Path Using a Sigma-Delta ADC

Figure 3.13 shows an example of this same application, using a $\Sigma\text{-}\Delta$ converter.

You can see that the $\Sigma\text{-}\Delta$ converter has a significant impact on the part count (between Figure 3.12 and 3.13) in a positive way. The converter in Figure 3.13 manages the acquisition of the photodiode signal and converts it, using a switched-capacitor input. The device immediately converts the signal to a digital representation with the $\Sigma\text{-}\Delta$ modulator and digital filter.

The design of $\Sigma\text{-}\Delta$ ADC in Figure 3.13 specifically targets transimpedance amplifiers. From this example, you can see the specialized nature of some $\Sigma\text{-}\Delta$ converters. This particular device (DDC114 from Texas Instruments) only targets photosensing applications.

Is the Sigma-Delta ADC Right for this Photodetection Application?

The sigma-delta ADC is well suited for this application. One advantage that this converter brings to the application is a reduced chip count.

The Right ADC for the Right Application

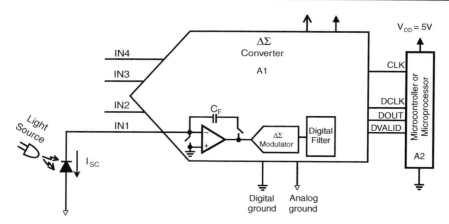

Figure 3.13: In this application circuit, the sigma-delta converter replaces the active anti-aliasing filter in Figure 3.12 with a passive R/C filter. The Σ–Δ converter also replaces the transimpedance amplifier (A2 and Rf). This new configuration has only two active devices plus the photosensor. In Figure 3.12, the part count is four active devices and eight passive parts.

Motor Control Solutions

The transition from mechanical, hydraulic and DC drives to higher-efficiency AC motor drives is triggering an increase in design updating in factory equipment. The traditional AC, motor-control system design used bulky and expensive current transformers for isolation and sophisticated digitizing systems to ensure the equipment performed in a well-behaved manner. The A/D conversion, in these AC induction motor systems, were capable of digitizing a larger full-scale range than was needed for the application and in packages that required more board space than the design could afford. The current transformers (also called Hall Effect Sensors) required an additional assembly step for board installation increasing the overall expense, which new designs don't easily tolerate.

New designs are approaching the isolation and digitization challenges presented by the motor control-application system from a different perspective in an attempt to lower costs and improve efficiency. The discrete, low-power, low-cost, differential input ADC maintains signal integrity in a potentially noisy environment. These sockets have 8-pin integrated devices that include the functions of common-mode rejection with the differential inputs, a gain stage, a sample/hold amplifier, and an ADC. All this functionality comes in package sizes as small as an MSOP. Digital optocouplers are replacing the bulky Hall Effect Sensors. These changes in the AC induction motor-sensing interface allows a more efficient, lower cost, smaller real-estate solution to the age old AC motor control design problem.

An AC induction motor control system block diagram is shown in Figure 3.14. In this motor-control servo-loop, a pulse width modulator (PWM) drive circuit switches the MOSFETs, which sends currents through the legs of the AC motor. A typical higher speed PWM motor

Chapter 3

control system uses fast switching, high power bipolar transistors, IGBTs, or MOSFETs to drive the three alternating currents through the AC motor. A SAR ADC digitizes the AC motor speed, position, and all three motor driving currents of the motor. The DSP chip receives the digitized information. The DSP engine evaluates the information received and initiates the next set of instructions to the PWM drive. The purpose of this local control system is to enhance the efficiency of the motor and to prevent catastrophic problems under all conditions, including temperature and load variations. The motor-control servo-loop implements this stability by adjusting the currents through the motor.

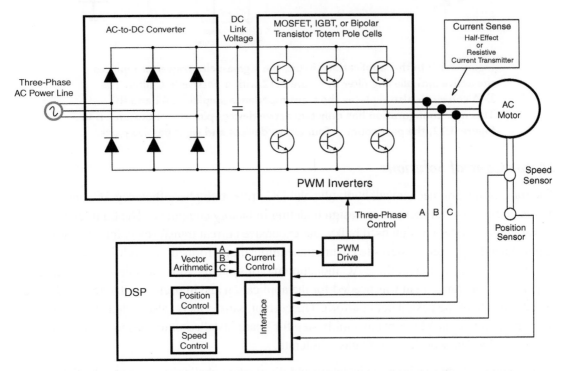

Figure 3.14: This is an AC motor control system configuration for three-phase control. The MOSFET, IGBT or bipolar transistor switches drive current through the AC motor. The "current sensors" sense this current. The ADC changes the signal from the current sensors into a digital code. The DSP engine then receives the digital code from the ADC. The DSP engine calculates appropriate PWM (pulse width modulator) instructions, again driving the totem pole power transistors into action.

The PWM motor control system drives high power bipolar transistors, IGBTs or power MOSFETs that in turn performed the "push-pull" action through the three legs of the AC motor. Since there are three legs and 360 degrees available in the clock cycle, each leg is switched 120 degrees apart from the other two. The timing between the signals and the magnitude of the currents provide critical information to the control loop when motor efficiency is a priority. The power devices that drive the motor are capable of handling currents up to

20A and voltage drops across the device several hundred volts to 600 V. An alternative design approach would be to sense the current in two of the three legs to the motor. Sensing systems that monitor only two legs at the motor are less expensive to implement, but are more susceptible to errors caused by grounding problems. A seemingly trivial circuit layout can be deceptive because of the higher than usual currents that are common in motor control circuits. Although, the circuits that monitor three legs are slightly more expensive because of the addition sensor channel, they are the more accurate of the two.

Switching speeds for the totem pole power transistors can typically range from 2 kHz up to 200 kHz. This places high demands on the driver circuit's ability to make a full-scale change, particularly with motors that operate with higher voltages.

The motor monitoring system is a critical link in the control system. This control system requires two different design approaches: The current sensing portion and the position/speed sensors. The current sensing circuitry simultaneously senses the three currents that are going to and from the motor. It consists of a sensing device, signal condition portion and isolation device.

The ADC requirements vary depending on the application requirements. An application where higher sensitivity is required, a 12-bit converter is appropriate. In applications where lower sensitivity is adequate, 8-bit to 10-bit converters are common. The controller receives the digital information from the ADC. The controller then assesses the condition of the motors. This processor(s) employ complex mathematics to perform vector transformations and PWM timing to finally regulate the motor to optimum efficiency. The processor(s) then feed back corrective action information to the PWM driver of the MOSFETs.

The motor controller's current sensor portion of the motor servo loop presents three interesting design challenges. The first challenge is that the motor servo loop requires that two to three signals are simultaneously sampled and converted to a digital representation within the limited cycle time of the motor drivers. High-end systems demand that the overall accuracy of these conversion systems is 12-bits. Simultaneous sampling can be an awkward problem to solve. Some designs use a front-end array of three sample/hold amplifiers and one high-speed 12-bit converter. The settling time of the sample/hold amplifiers and the throughput rate of the one converter limits the throughput rate of this configuration. In order to digitize a complete picture of the motor at a particular instance in time, the converter must digitize two or three separate voltages. Each A/D conversion added to this digitization process adds additional time to the throughput of the sensor stage.

A second issue that presents design challenges in the sensor stage circuit is the conversion of the sensor current to a good system-level signal. An example of the MOSFET switching section of the sensor stage circuit is shown in Figure 3.15b. The MOSFET devices are configured in a totem pole arrangement using one leg of the motor as the output load. When the totem pole switches from high to low (or low to high), the current output (I_O) magnitude and direction changes. The motor sensor is positioned at the output of the totem pole switch in order to capture the changes in I_O. The sensor in these circuits has negligible effects on the operation of the motor. As an example, a small value, power resistor (R_{SENSE}), as shown in Figure 3.15b can sense the pertinent information with very little voltage drop and power loss. In contrast, the voltage change at the output of the totem pole can be rather large. So, the trick is to sense the small voltage drop across R_{SENSE} while rejecting the large voltage excursions of the output of the MOSFET totem pole. Once the large common-mode interference from the motor is rejected, the signal is gained to a usable level for the ADC.

Third, if the signal is in the analog domain, electromagnetic interference (EMI) becomes an issue. The changes in the current magnitude and direction through the motor legs are an analog signal. Careful layout and design will save headaches later in the design. If the signal is in the analog domain, it should be a differential signal. The analog portion of the motor sensing circuitry can most easily reject the noise present if the system has differential signals throughout. The best scenario would be to digitize the signal as soon as possible.

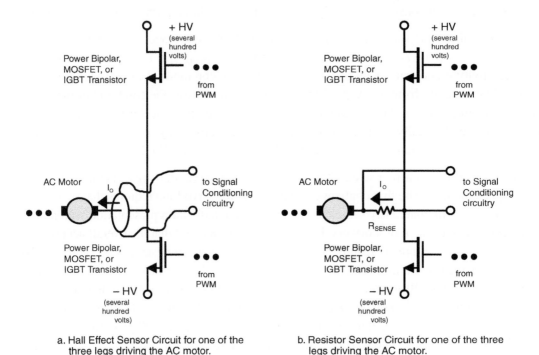

a. Hall Effect Sensor Circuit for one of the three legs driving the AC motor.

b. Resistor Sensor Circuit for one of the three legs driving the AC motor.

Figure 3.15: The Hall Effect Sensor (a) magnetically senses the changes in current initiated by the PWM inverter in one of the three driver legs to the motor. The transfer function of a Hall Effect Sensor is current-to-current. The conditioning circuitry that follows is required to convert the output current of the Hall Effect Sensor to a voltage. A sensing small value, power resistor (b.) can replace the Hall Effect Sensor. This resistor senses the changes in current to and from the AC motor and immediately converts it to a voltage.

The Right ADC for the Right Application

The previously mentioned transition from DC drives to higher efficiency AC motors has been motivated by lowered costs in the AC motor control system coupled with the better power efficiency, which has always been an AC motor characteristic. The target areas for cost reduction have been with the sensing circuitry, the ADC(s) and the microcontroller. Cost reduction of the sensing circuitry and the A/D conversion portion of the circuit have been largely driven by technology strides in terms of higher integration and reduced chip layout dimensions. The combination of these two phenomena has led to reduced chip sizes and a lower chip count for the application. In both cases, the end user enjoys these cost savings. Figure 3.16 shows one possible design approach for the current sensing circuitry and ADC conversion portion.

In the sensing circuit shown in Figure 3.16, the ADCs are directly interfaced to the sensing resistors, R_{SENSE}. The input ranges of the converters are programmable by the voltage reference to the devices, ensuring that the full dynamic range of the ADCs is used. With the ADC in this circuit, the voltage reference input is 200 mV. This fully eliminates the need for

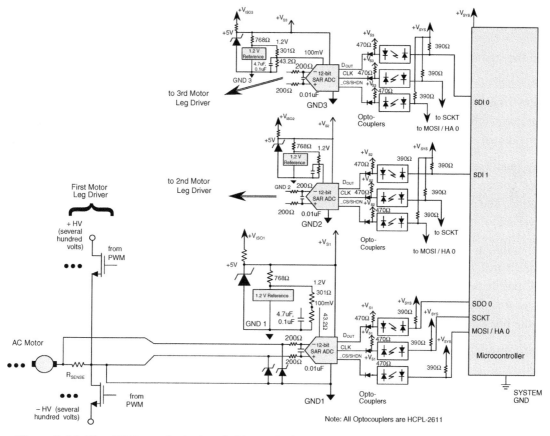

Figure 3.16: The ADC, on the isolated side of the isolation barrier, is part of the motor control sensing circuitry. The optocouplers transmit a digital signal across the barrier. Since an ADC digitizes the signal at the sensor, this circuit approach allows for better noise immunity.

Chapter 3

an analog gain stage. Although the ADC is a 12-bit converter, the effective accuracy of the device with a 100 mV reference is 11-bits. The ADC differentially senses the sense resistor, now, directly. The differential inputs of the ADCs assist in maintaining good signal integrity through the completion of the digitizing process.

The chosen method of isolation for this configuration is a digital optocoupler instead of a Hall Effect Sensor. The digital section of the ADC is isolated from the processor by way of three optocouplers. The digital output of each of the three converters is sent through an array of three optocouplers, then to the microcontroller. All nine digital optocouplers in this circuit must have excellent transient immunity. The throughput rate of the optocouplers must also be fast enough for the clock input and the data output of the ADC. In this application, all three converters are triggered at the same time. This meets the design requirements of three simultaneous sampling channels, thereby reducing the throughput rate of the sensor circuitry.

Conclusion

So, here we are at the end of a chapter where we have discussed the ADCs for your circuits and how they fit with various analog signals. In particular, we concentrated on the SAR ADC and the sigma-delta ADC.

It turns out that through this discussion, we were forced to look at the entire signal chain. You will find that this is a requirement when you design with an ADC in your circuit. It is one thing to find the "perfect" converter, and another thing to find the right converter for your signal chain.

As I mentioned in Chapter 2, the ADC is almost an afterthought. Your real task is to figure out what you need done in your signal chain and then pick the right converter, not the other way around. The punch line is if the ADC reports incorrect data, the controller, or processor will never know, unless you write code that identifies errors in your analog system.

Generally, in terms of selecting the right product for the applications, SAR converters will require an anti-aliasing filter of 2nd to 5th order, where the sigma-delta ADC only requires an R/C pair to perform this kind of filtering. SAR converters have lower resolutions but they are faster than the sigma-delta ADC. If the signal is slow enough, the sigma-delta ADC can easily fit the bill while eliminating a lot of the analog front end.

The converter should fit the application. Remember, with the ADC the best defense is a good, well-informed offense.

Chapter 3 References

"Get Maximum Accuracy from Temperature Sensors," Steele, Jerry, *Electronic Design*, August 19, 1996, p. 99.

"Understanding and Using PRTD Technology, Part I: History, Principles, and Designs," Sulciner, James, *Sensors*, August, 1996, p. 10.

"Understanding and Using PRTD Technology, Part II: Selection," Sulciner, James, Sensors, September, 1996, p. 43.

1994 Application Seminar, Burr-Brown, Chapter 1.

"OMEGA Temperature Measurement Handbook & Encyclopedia," Omega Engineering Inc., Stamford, CT.; 1996.

"Practical Temperature Measurements," Omega Catalog, p. Z-11.

"Evaluating Thin Film RTD Stability," *Sensors*, Hyde, Darrell, Oct. 1997, p. 79.

"Refresher on Resistance Temperature Devices," Madden, J.R., *Sensors*, Sept. 1997, p. 66.

"Producing Higher Accuracy From SPRTs (Standard Platinum Resistance Thermometer)," MEASUREMENT & CONTROL, Li, Xumo, June, 1996, p. 118.

"Pressure Sensors," Tandeske, Duane, Marcel Dekker, Inc., 1991.

"Anti-Aliasing Analog Filters for Data Acquisition Systems," Baker, Bonnie C., AN699, Microchip Technology Inc.

"Single Supply Temperature Sensing with Thermocouples," Baker, Bonnie C., AN684, Microchip Technology, Inc.

Discrete-time Signal Processing, Oppenheim, Schafer, Prentice Hall, Inc., 1989.

"Number of Bits vs. LSB Errors," Baker, Bonnie C., *EDN Magazine*, July 8, 2004.

CHAPTER 4

Do I Filter Now, Later or Never?

CHAPTER 4

Do I Filter Now, Later or Never?

Have you ever needed a low-pass filter (LPF) or high-pass filter (HPF) in your circuit and wanted to place it in the signal path? Prior to the introduction of controllers and processors, analog circuits were used to implement all types of filters. With this type of system, you needed to think ahead and work up a "pencil design" before going to the breadboard. If you cut corners you would probably end up disassembling the circuit and rebuilding it, hoping to get it right the next time.

Then came the digital filter. This type of filter implementation is capable of duplicating the frequency response of any analog filter in the digital domain. A major advantage of the digital filter is that you can adjust it painlessly with firmware. This sounds like it is too good to be true, and it is. Way too good to be true.

There are times when you should build the filter with analog hardware, and times where it is appropriate to implement it with a controller or processor in firmware. An analog LPF should be in every circuit where there is an analog-to-digital conversion. This is true whether the analog-to-digital converter (ADC) is a successive approximation register (SAR), sigma-delta (Σ–Δ), pipeline, dual slope or any other type of converter you may contrive. The placement of this type of filter in the circuit must always be on the analog side, in front of the converter as shown in Figure 4.1.

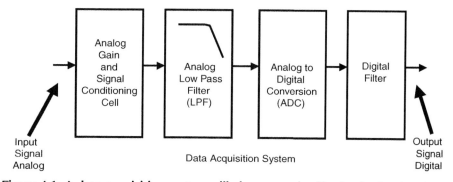

Figure 4.1: A data acquisition system will always require filtering in the signal path to remove unwanted signals. The analog filtering is required in order to get rid of higher frequency noise that is outside half of the sampling frequency of the converter. You use the digital filter to reduce lower in-band, frequency noise.

Chapter 4

The system shown in Figure 4.1 includes an analog filter and a digital filter. When you use an analog filter, you include it prior to the ADC. In contrast, you include a digital filter after the analog-to-digital conversion has occurred. It is obvious why you would implement these two filters at these particular points. However, the ramifications of these positions are not quite so obvious. There are a number of system differences when the filtering function is in the analog domain, rather than the digital domain. You should be aware of these differences.

Why do you need an analog LPF? Remember that every analog signal has high and low frequency noise. This is true whether you acknowledge this fact or not. The reason that an LPF is required goes back to the Nyquist theorem, which illustrates that any signal (or noise), at any frequency has a chance of being accurately converted or possibly contaminated.

Any signal that passes through the ADC has a magnitude associated with it. The ADC reliably converts this signal magnitude as long as the signal frequency is below the converter's input-stage bandwidth. Although the magnitude is preserved, the same is not true for the signal's frequency. The signal frequencies above ½ of the sampling frequency of the ADC are contaminated to the point where you won't be able to tell the difference between the digital representation of input signals below ½ of the sampling rate and those ½ above the sampling rate. You might already know this as signal aliasing, per the Nyquist theorem. You'll find more details on this theorem later on in this chapter. As long as you know how the Nyquist theorem works, you will be able to see how unwanted noise and signals can become a permanent part of the digital signal. If you don't see this, we will go into more detail later on. But, once this contamination has occurred, there is no going back and undoing it.

A low-pass analog filter removes superimposed higher frequency noise from the analog signal before it reaches the ADC. This also includes extraneous noise peaks. Digital filtering cannot eliminate these peaks riding on the analog signal. Therefore, noise peaks riding on signals near full-scale have the potential to saturate the analog modulator of the ADC. This is true, even when the average value of the signal is within limits. With an LPF that is prior to the ADC, the task of successfully achieving high-resolution is placed squarely on the analog circuit design and converter.

In contrast, a digital filter, by definition uses oversampling and averaging techniques to reduce in-band noise. Since digital filtering occurs after the A/D conversion process, it can remove noise injected during the conversion process (like quantization noise). Analog filtering cannot do this. Also, the digital filter is programmable far more readily than an analog filter. Depending on the digital filter design, this gives you the capability of programming the cut-off frequency and output data rates.

Implementing an HPF and a second LPF could be a digital filter function or analog. The advantage of building these filters in the controller or processor is that you can implement a large variety of, easy to adjust filters. The variety includes the analog style filters, such as Butter-worth, Bessel, Chebyshev, or Elliptic, however you can also implement digital filters such as the

Do I Filter Now, Later or Never?

FIR filter, an IIR filter or fast Fourier transform (FFT). By implementing a FIR filter, you can significantly reduce in-band noise as discussed in Chapter 2 (Sigma-Delta ADC section). You cannot achieve this reduction of in-band noise with an analog filter. An FFT easily removes unwanted frequencies digitally. In an analog circuit this can be done, but it is hardware intensive.

With digital filter designs on the horizon, the analog filter looks unattractive. However, they still have a place in the signal path, as does the digital filter. For now the good news is that, the digital filter has taken us to the next level of performance, precision and cost reduction.

In the first part of this chapter, we define the analog, low-pass filter design parameters. These definitions apply to passive and active filters. You can build passive filters with resistors, capacitors and/or inductors. The components of active filters are operational amplifiers, capacitors, and resistors. We will discuss the frequency characteristics of a low-pass filter with some reference to specific filter designs.

In the second section, the discussion will move to an in-depth look at low-pass filter designs. At this point, we will define the approximation types, such as Butterworth, Bessel and Chebyshev. The next portion of this chapter will discuss techniques used when determining the appropriate filter design parameters of an active anti-aliasing filter. In this section, the discussion will turn to the aliasing theory. Following the topic of aliasing theory, I will show examples of passive and active low-pass filter circuits.

In the third section of this chapter, we will talk about applications. We will design the appropriate filter for an RTD temperature sensing circuit, a circuit with an analog multiplexer before the filter, and a precision light sensing circuit.

Key Low-Pass Analog Filter Design Parameters

The frequency domain specifications of a low-pass analog filter use four parameters (see Figure 4.2). These four parameters are: $f_{CUT-OFF}$, f_{STOP}, A_{MAX} and M. The cut-off frequency ($f_{CUT-OFF}$) of a low-pass filter defines the –3 dB point for a Butterworth and Bessel filter. It also describes the frequency at which the filter response leaves the error band for the Chebyshev.

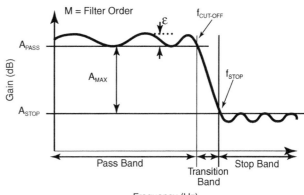

Figure 4.2: The key analog filter design parameters include the –3dB cut-off frequency of the filter ($f_{CUT-OFF}$), the frequency at which a minimum gain is acceptable (f_{STOP}) and the number of poles (M) implemented with the filter.

Chapter 4

The frequency span from DC to the cut-off frequency is the pass-band region. The magnitude of the response in the pass band is A_{PASS} in Figure 4.2. The response in the pass band can be flat with no ripple, as it is with a Butterworth or Bessel filter. Conversely, a Chebyshev filter has a ripple up to the cut-off frequency. The magnitude of the ripple error of this filter is ε.

By definition, a low-pass filter passes lower frequencies up to the cut-off frequency and attenuates the higher frequencies that are above the cut-off frequency. An important parameter is the filter system gain, A_{MAX}. A_{MAX} is the difference between the gain in the pass-band region and the maximum gain in the stop-band region or $A_{MAX} = A_{PASS} - A_{STOP}$.

If there is ripple in the pass band of the filter response, the magnitude of the gain of the pass-band (A_{PASS}) is at the bottom of the ripple. The stop-band frequency, f_{STOP}, is the frequency at which a minimum attenuation is reached. It is possible that the stop band has a ripple. The minimum gain (A_{STOP}) of this ripple is the highest peak of the stop band. If the filter response has no stop-band ripple, as with the Butterworth or Bessel filter, the application requirements determine the stop-band frequency. For example, you as a designer may calculate that the maximum amount of higher frequency noise has an amplitude of –40 dB riding on your signal. If this is the case, you will define the stop-band frequency at the point where the filter has attenuated 40 dB from the DC value.

As the response of the filter goes beyond the cut-off frequency, it falls through the transition band to the stop band region. The bandwidth of the transition band is determined by the filter design (Butterworth, Bessel, Chebyshev, and so forth) and the order (M) of the filter. The number of poles in the transfer function determines the filter order. For instance, if a filter has three poles in its transfer function, it is a 3rd order filter.

Generally, the transition band will become smaller when you use more poles to implement the filter design. Figure 4.3 illustrates a Butterworth filter. Ideally, a low-pass, anti-aliasing filter should perform with a "brick wall" style of response, where the transition band is designed to

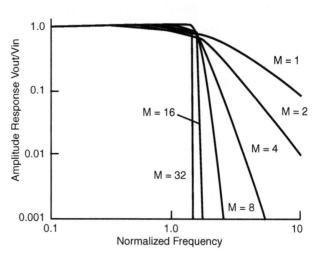

Figure 4.3: If you increase the number of poles in a Butterworth low-pass filter design, you get a sharper roll-off through the transition region. Each response is dependent on the number of poles or the order (M) of the filter.

be as small as possible. Practically speaking, this may not be the best approach for an anti-aliasing solution. With an active filter design, every two poles require an operational amplifier. For instance, if you design a 32nd order filter, you will need 16 operational amplifiers, 32 capacitors and up to 48 resistors to implement the circuit. Additionally, each amplifier would contribute offset and noise errors into the pass-band region of the response.

Figure 4.4 shows a low-pass filter response in the time domain. The input signal for this type of filter response is a square wave. This figure shows five, low-pass filter characteristics. You will use these five characteristics to further define and design your filter. The first characteristic is the propagation delay time. The phase margin of your filter at the cut-off frequency determines the propagation delay time. The propagation time is the time between the initiation of the rising edge of the input square wave and the response of the filter response at its output. Generally, if the cut-off frequency is low, the propagation delay time will be greater than if the cut-off frequency is high.

The second characteristic is the slew rate (SR) of the output signal attempting to match the input signal. This parameter's unit of measure is change of voltage with time or V/sec. Once the filter successfully goes to its final destination, the signal can exhibit some overshoot. This overshoot settles to the final value that has an acceptable error (ε). The unit of measure for overshoot is seconds. The unit of measure for the final value error is percentage of full-scale.

The settling time (t_S) is a combination of the time required for the filter to slew, plus the overshoot activity until the output of the filter finally settles to an acceptable error band.

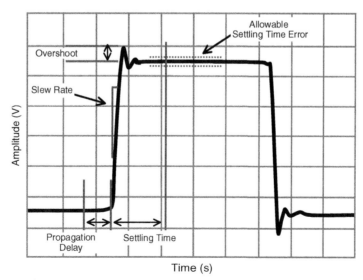

Figure 4.4: The time domain plot of a filter shows the output of the filter as it reacts to a square-wave, input signal. The key characteristics of this response are propagation delay, slew rate (SR), overshoot, settling time (ts), and settling time error (ε).

Chapter 4

Generally, it is a good rule of thumb to design active filters with the number of poles ranging from one to five. If the number of poles in the filter is higher, it may be difficult to build a stable filter and it can become cost prohibitive. We will discuss strategies on how to work around these limitations in the "Anti-Aliasing Filter Theory" section of this chapter.

Analog Filter Approximation Types

The more popular filter approximation types are the Butterworth, Bessel and Chebyshev. If you examine the filter amplitude versus frequency domain response and their amplitude versus time domain response, you can identify any filter. Other filter types not discussed in this chapter include Inverse Chebyshev, Elliptic and Cauer designs. I'll try to save that for another book.

Butterworth Filter

The Butterworth filter is by far the most popular filter design. The transfer function of a Butterworth filter consists of all poles and no zeros and is equal to:

$V_{OUT}/V_{IN} = G/(a_0 s^n + a_1 s^{n-1} + a_2 s^{n-2} ... a_{n-1} s^2 + a_n s + 1)$
where G is equal to the gain of the system.

Table 4.1 lists the denominator coefficients for Butterworth designs. Although the order of a Butterworth filter design theoretically can be infinite, this table only lists coefficients up to a 5th order filter.

Table 4.1: This table lists the coefficients versus filter for 2nd, 3rd, 4th and 5th order Butterworth designs.

M	a_0	a_1	a_2	a_3	a_4
2	1.0	1.4142136			
3	1.0	2.0	2.0		
4	1.0	2.6131259	3.4142136	2.6131259	
5	1.0	3.2360680	5.2360680	5.2360680	3.2360680

As shown in Figure 4.5, the pass-band portion of the Butterworth filter curve is flat. The technical term for this characteristic is *maximally flat*. You will see later that the rate of attenuation in transition band is better than Bessel, but not as good as the Chebyshev filter. There is no ringing in stop band. This helps with your accuracy if you are using higher resolution ADCs in the signal chain.

Figure 4.6 shows the step response of the Butterworth low-pass filter. In this figure, you can see that the Butterworth filter has some overshoot and ringing in the time domain. You will see later on that a Butterworth filter rings less than the Chebyshev filter but more than the Bessel filter. If the filter order is higher, this overshoot will also be higher.

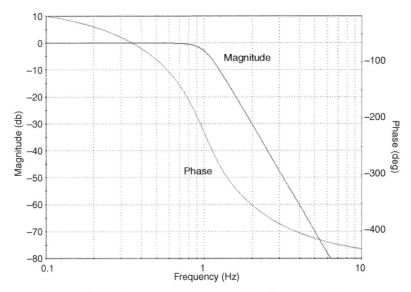

Figure 4.5: The frequency response of the Butterworth low-pass filter is maximally flat in the pass band. This plot shows the response of a 5th order Butterworth filter.

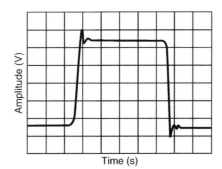

Figure 4.6: The step response of a 5th order, low-pass Butterworth filter has some ringing, but it eventually will settle after time. The amplitude of the ringing is dependent on the order (or number of poles) of the filter. The settling time is dependent on the filter's corner frequency and order. Generally speaking, the higher the order of the filter the higher the overshoot will be and the settling time will be longer.

If you are using this filter after a multiplexer, the filter settling time needs to be considered. This is because there is a possibility that the multiplexer output-signal change going into the filter can be a step response as you change from one channel to another.

Chapter 4

Chebyshev Filter

The transfer function of the Chebyshev filter is only similar to the Butterworth filter in that it has all poles and no zeros with a transfer function of:

$$V_{OUT}/V_{IN} = G/(a_0 + a_1s + a_2s^2 + \ldots a_{n-1}s^{n-1} + s_n)$$

Its frequency behavior has a ripple (Figure 4.7) in the pass band; the pole placement in the circuit design determines this ripple. Figure 4.2 shows the magnitude of the ripple as ε. In general, an increase in ripple magnitude will lessen the width of the transition band.

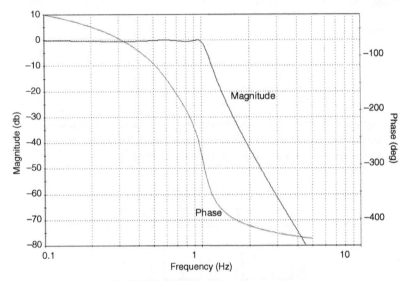

Figure 4.7: This is the frequency response of a 5th order, low-pass Chebyshev filter. This filter design has a 0.5 dB ripple in the pass band region. After the cut-off frequency, the attenuation of this filter is faster than the Butterworth or Bessel filter.

Table 4.2 gives the denominator coefficients of a 0.5 dB ripple Chebyshev design. Although the order of a Chebyshev filter design theoretically can be infinite, this table only lists coefficients up to a 5th order filter.

Table 4.2: Coefficients versus filter order for 2nd, 3rd, 4th and 5th order Chebyshev filters with a 0.5 db ripple.

M	a_0	a_1	a_2	a_3	a_4
2	1.516203	1.425625			
3	0.715694	1.534895	1.252913		
4	0.379051	1.025455	1.716866	1.197386	
5	0.178923	0.752518	1.309575	1.937367	1.172491

The ripple magnitude of ε (Figure 4.2) theoretically can be as large or as small as you want it to be. But the amount of error you can tolerate through the pass band or the speed of the frequency-response, and attenuation in the transition band limits your choice. In general terms, a high-ripple magnitude will provide more error in the pass-band region, but a faster attenuation in the transition band.

The rate of attenuation in the transition band is steeper than the Butterworth and Bessel filters. For instance, a 5th order Butterworth filter is required if it is to meet the transition bandwidth of a 3rd order Chebyshev filter with a 0.5 dB ripple. Although there is ringing in the pass-band region with this filter, the stop band is void of ringing.

The step response (Figure 4.8) of a 5th order Chebyshev low-pass filter with a 0.5 dB ripple has a fair degree of overshoot and ringing.

The overshoot and ringing phenomena is a result of the phase response in the frequency domain. If you will remember, the Fourier analysis of a step response (or square wave) shows that you can construct a square wave by adding odd harmonic sinusoidal signals. If you re-examine the magnitude versus frequency plot in Figure 4.7, you will notice that the ratio of frequency to phase is not constant from lower frequencies to f_{STOP}. Therefore, the higher frequencies from the step input arrive at the output of the filter before the lower frequencies. This distortion is sometimes called *group delay* and is seen as ringing at the output of the filter and can be calculated with (delta phase/delta f)/360 (in seconds).

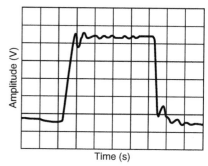

Figure 4.8: This is the step response of a 0.5 dB, low-pass, 5th order Chebyshev filter. With this filter the ringing is longer lasting than you will find with the Butterworth or Bessel filter.

Bessel Filter

Once again, the transfer function of the Bessel filter has only poles and no zeros. The Butterworth design is maximally flat in the pass-band region. The Bessel filter produces a constant time delay with respect to frequency over a large range of frequency. Mathematically, this relationship is expressed as:

$C = -\Delta\theta \times \Delta f$

where

C is a constant,
θ is the phase in degrees, and
f is frequency in Hz

Chapter 4

Alternatively, this relationship can be expressed in degrees per radian as:

$$C = -\Delta\theta / \Delta\omega$$

where

C is a constant,
θ is the phase in degrees, and
ω is in radians.

The transfer function for the Bessel filter is:

$$V_{OUT}/V_{IN} = G/(a_0 + a_1 s + a_2 s^2 + ... a_{n-1} s^{n-1} + s_n)$$

Table 4.3 shows the denominator coefficients for Bessel filters. Although the order of a Bessel filter design theoretically can be infinite, this table only lists coefficients up to a 5th order filter.

Table 4.3: Coefficients versus filter order for 2nd, 3rd, 4th and 5th order Bessel designs.

M	a_0	a_1	a_2	a_3	a_4
2	3	3			
3	15	15	6		
4	105	105	45	10	
5	945	945	420	105	15

The Bessel filter has a flat magnitude response in the pass band (Figure 4.9). Following the pass band, the rate of attenuation in the transition band is slower than the Butterworth or Chebyshev. And finally, there is no ringing in the stop band.

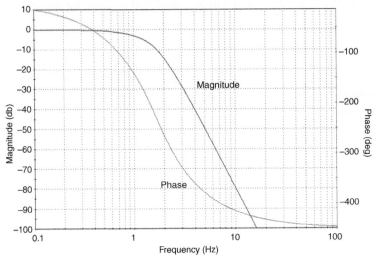

Figure 4.9: This is a 5th order, low-pass, Bessel filter response over frequency. The ratio of phase over frequency is relatively constant.

The design in Figure 4.10 has the best step response of the three designs in this chapter, with very little overshoot or ringing. This response is easy to predict in the frequency domain by evaluating the ratio of change in frequency with change in phase. In the case of the Bessel filter, this ratio is nearly constant.

Anti-Aliasing Filter Theory

ADCs usually operate with a constant sampling frequency when digitizing analog signals. By using a sampling frequency (f_S), typically called the *Nyquist frequency*, you can reliably digitize all input signals with frequencies below $f_S/2$. If there is a portion of the input signal that resides in the frequency domain above

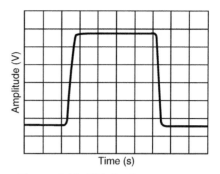

Figure 4.10: This is the step response of a 5th order, low-pass Bessel filter. As you can see in this response, there is virtually no ringing at the top or bottom of the step; only a slight time delay.

$f_S/2$, that portion will fold back into the bandwidth of interest with the amplitude preserved. This phenomenon makes it impossible to tell the difference between a signal from the lower frequencies (below $f_S/2$) and higher frequencies (above $f_S/2$). Figure 4.11 shows this aliasing or fold-back phenomenon in the frequency domain.

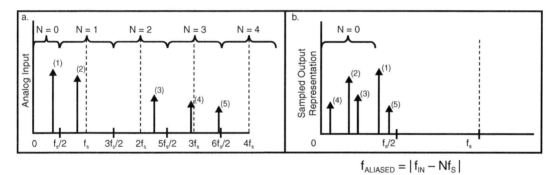

$$f_{ALIASED} = |f_{IN} - Nf_S|$$

Figure 4.11: A system that is sampling frequency at f_s (a) will identify signals with frequencies below $f_s/2$ as well as above. Input signals below $f_s/2$ will be reliably digitized while signals above $f_s/2$ will be folded back (b) and appear as lower frequencies in the digital output.

In both parts of this figure, the x-axis identifies the frequency of the sampling system, f_S. The left portion of Figure 4.11 shows five segments of the frequency band. Segment $N = 0$ spans from DC to one-half of the sampling frequency. In this bandwidth, the sampling system will reliably record the frequency content and magnitude of an analog input signal. In the segments where $N > 0$ the frequency content of the analog signal will be recorded by the digitizing system in the bandwidth of the segment $N = 0$ on the left side of this figure. Mathematically, the higher frequencies (2, 3, 4, 5) will fold back with the following equation:

$$f_{ALIASED} = |f_{IN} - Nf_S|$$

Chapter 4

This fold-back phenomenon is illustrated in Figure 4.11b where signals 2, 3, 4 and 5 appear at lower frequencies than in Figure 4.11a.

This is how this formula plays out. Let the sampling frequency, (f_S), of the system be equal to 100 kHz and the frequency content of:

$f_{IN}(1) = 41$ kHz

$f_{IN}(2) = 82$ kHz

$f_{IN}(3) = 219$ kHz

$f_{IN}(4) = 294$ kHz

$f_{IN}(5) = 347$ kHz

The sampled output contains accurate amplitude information of all of these input signals. However, four of the signals will fold back into the frequency range of DC to $f_S/2$ or DC to 50 kHz. By using the equation $f_{OUT} = |f_{IN} - Nf_S|$, the frequencies of the input signals are transformed to:

$f_{OUT}(1) = |41 \text{ kHz} - 0 \times 100 \text{ kHz}| = 41$ kHz

$f_{OUT}(2) = |82 \text{ kHz} - 1 \times 100 \text{ kHz}| = 18$ kHz

$f_{OUT}(3) = |219 \text{ kHz} - 2 \times 100 \text{ kHz}| = 19$ kHz

$f_{OUT}(4) = |294 \text{ kHz} - 3 \times 100 \text{ kHz}| = 6$ kHz

$f_{OUT}(5) = |347 \text{ kHz} - 4 \times 100 \text{ kHz}| = 53$ kHz

Note that all of the signal frequencies above $f_S/2$ have a change in frequency at the output of the converter. If you choose to remove these signals after an A/D conversion, it will be very difficult. This is true, particularly if you really don't know all of the details about the frequency content of the signal at the input.

An analog, low-pass filter prior to the input of the ADC eliminates or significantly reduces aliased noise. Figure 4.12 illustrates this concept. In this diagram, the low-pass filter attenuates the second portion of the input signal at frequency (2). Therefore, this signal will not be aliased into the final sampled output. There are two regions of the analog, low-pass filter illustrated in Figure 4.12. The region to the left is within the bandwidth of DC to $f_S/2$. The second region, which is shaded, illustrates the transition band of the filter. Since this region is greater than $f_S/2$, signals within this frequency band will be aliased into the output of the sampling system.

Moving the corner frequency of the filter lower than $f_S/2$ or increasing the order of the filter can minimize the affects of this error. In both cases, the minimum gain of the filter, A_{STOP}, at $f_S/2$ should be less than the noise riding on your signal.

Do I Filter Now, Later or Never?

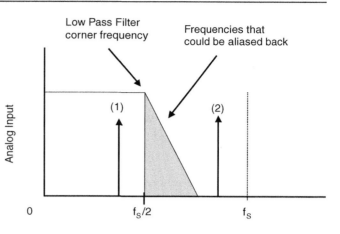

Figure 4.12: A low-pass, analog filter prior to the ADC of a sampling system attenuates higher frequencies. The anti-aliasing filter prevents the higher frequencies from being sampled or folded back into the digital signal.

Analog Filter Realization

Traditionally, passive devices were used to build low-pass filters—that is, resistors and capacitors. High-pass or bandpass filters needed inductors. Active filter designs were realizable or in these components. However, the cost of operational amplifiers was prohibitive. Filter designs still use passive filters when a single-pole filter is required or when the bandwidth of the filter is above the gain bandwidth product of operational amplifiers on the market. Even with these two exceptions, filter realization is predominately implemented with operational amplifiers, capacitors and resistors.

Passive Filters

Passive, low-pass filters use resistors and capacitors. Figure 4.13 shows the realization of a single-pole, low-pass filter.

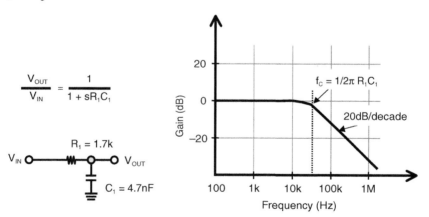

Figure 4.13: The passive, low-pass, analog filter uses a resistor and capacitor. The DC input and output impedance of this type of filter implementation is equal to R_1. The passive implementation is less expensive; however, the input and output resistance is high. As a result, this configuration usually needs one or two buffers.

Chapter 4

The output impedance of a passive, low-pass filter is relatively high when compared to the active filter realization. With the active filter realization, the output impedance is equal to the output resistance of the amplifier. For example, a 1 kHz low-pass filter, which uses a 0.1 µF capacitor in the design, would need a 1.59 kΩ resistor. This value of resistor could create an undesirable voltage drop or make impedance matching difficult. As a result, you need to use some degree of care, typically when implementing passive filters. Single-pole operational amplifier filters have the added benefit of "isolating" the high impedance of the filter from the following circuitry.

It is very common to use single-pole, low-pass filters at the input of a sigma-delta ADC or the input of a switched capacitor filter. In this case, the high output impedance of the filter usually does not interfere with the conversion process unless you are using really high resistors. But if you use high value resistors, you will be contributing noise to the lower frequency bands. For more information about resistor noise, refer to Chapter 10.

Active Filters

A 2nd order active filter uses a combination of one amplifier, one to three resistors, and one to two capacitors to implement one or two poles. The active filter offers the advantage of providing "isolation" between stages. This is possible by taking advantage of the high input impedance and low output impedance of the operational amplifier. In all cases, the number of capacitors at the input and in the feedback loop of the amplifier determines the filter order.

These 1st order or 2nd order filters can be put in series, making higher order filters easy to design. For example, if two 2nd order active low-pass filters are in series, the total order of the filter is 4th order. Since these two 2nd order filters are both active, the input impedance of each stage is usually high and the output impedance is low. As a result, the two stages don't interact. This makes it easier to design each stage successfully.

Single-Pole Filter

The frequency response of the single-pole, active filter is identical to a single-pole passive filter. Figure 4.14 illustrates examples of the realization of single-pole active filters.

The input impedance of the filter in Figure 4.14a is very high, and equivalent to the input impedance of the amplifier. For CMOS amplifiers, this is in the range of 10^{13} Ω. The output impedance of this circuit is very low, compared to the passive filter. It typically falls in the range of tens to a few hundred ohms. The input impedance of the filter in Figure 4.14b is equal to R_1. The output impedance of this circuit is the same as the output impedance of the circuit in Figure 4.14a. These two circuits provide these benefits while producing the same transfer function of the passive, low-pass filter.

Do I Filter Now, Later or Never?

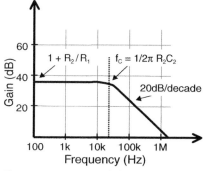

a. Single pole, non-inverting active filter

b. Single pole, inverting active filter

c. Frequency response of single pole active filter

Figure 4.14: You can build a 1st order filter with an operational amplifier in combination with two resistors and one capacitor. Figure 4.13 shows that the design of this type of low-pass filter can exclusively use passive elements, or with an amplifier and R/C pair. The noninverting and inverting active implementations have low output resistance. Additionally, the noninverting implementation has very high input impedance.

Double-Pole, Voltage-Controlled Voltage Source or Sallen-Key Filter

The double-pole, voltage-controlled voltage source is known as the Sallen-Key filter realization. The DC gain of this filter is positive. Figure 4.15 shows a 2nd order, Sallen-Key realization. This filter has a DC gain of 1 V/V. The poles of this filter are determined by the resistive and capacitive values of R_1, R_2, C_1 and C_2.

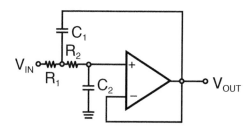

Figure 4.15: The Sallen-Key filter implementation with a DC gain is equal to 1 V/V. The signal that passes through this filter is not inverted.

Chapter 4

Double-Pole Multiple Feedback

Figure 4.16 shows the double-pole, multiple feedback realization of a 2nd order low-pass filter. Sometimes, the name *multiple feedback filter* identifies this filter type. The DC gain of this filter inverts the signal and is equal to the ratio of R_1 and R_2. The values of R1, R2, R3, C5 and C6 determine the frequency of the poles of this filter.

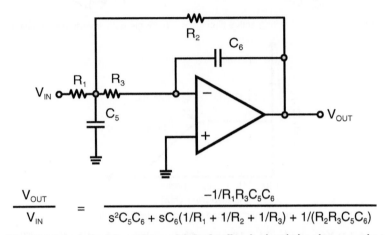

$$\frac{V_{OUT}}{V_{IN}} = \frac{-1/R_1R_3C_5C_6}{s^2C_5C_6 + sC_6(1/R_1 + 1/R_2 + 1/R_3) + 1/(R_2R_3C_5C_6)}$$

Figure 4.16: A double-pole, multiple feedback circuit implementation uses three resistors and two capacitors to implement a 2nd order analog filter. DC gain is equal to $-R_2/R_1$.

How to Pick Your Operational Amplifier

The task of selecting the correct single-supply operational amplifier for an active low-pass filter circuit can appear overwhelming, especially when reading an op amp data sheet and viewing all of the specifications. For example, the number of AC and DC electrical specifications in Microchip's 5 MHz, single-supply, MCP6281/2/3/4 data sheet is 24. But in reality, there are only two important specifications that you should initially consider when selecting an op amp for your active, low-pass filter. Once you have chosen your amplifier, based on these two specifications, you should consider two additional specifications before reaching your final decision.

The two key specifications that you should initially consider when designing with either the Sallen-Key or multiple feedback topologies is *gain bandwidth product* and *slew rate*. Prior to the selection of the op amp, you need to determine the filter cut-off frequency (f_C).

Since you have already defined your cut-off frequency, selecting an amplifier with the right bandwidth is easy. The closed-loop bandwidth of the amplifier must be at least 100 times higher than the cut-off frequency of the filter. If you are using the Sallen-Key configuration and your filter gain is +1 V/V, use the following formula for this type of filter:

$GBWP \geq 100 \times f_C$,
where GBWP is equal to the gain bandwidth product of the amplifier.

If you are using the multiple feedback configuration, use the following formula.

$GBWP \geq 100 \times (-G_{CLI} + 1) \times f_C$,
where G_{CLI} is equal to the inverting gain of your closed-loop system.

In addition to paying attention to the bandwidth of your amplifier, the slew rate should be evaluated in order to ensure that your filter does not create signal distortions. Internal currents and capacitances determine the slew rate of an amplifier. When you send large signals through the amplifier, the appropriate currents charge these internal capacitors. The speed of this charge is dependent on the value of the amplifier's internal resistances, capacitances and currents. In order to ensure that your active filter does not enter into a slew condition, you need to select an amplifier such that:

Slew rate $\geq (2 \times \pi \times V_{OUT\ P-P} \times f_C)$,
where $V_{OUT\ P-P}$ is the expected peak-to-peak output voltage swing below f_C.

There are two, secondary specifications that affect your filter circuit. These are: Input common-mode voltage range (V_{CMR}) for the Sallen-Key circuit, and input bias current (I_B). In the Sallen-Key configuration, V_{CMR} will limit the range of your input signal. The other 2nd order specification to consider is the I_B. This specification describes the amount of current going in or out of the input pins of the amplifier. If you are using the Sallen-Key filter configuration, as shown in Figure 4.15, the input bias current of the amplifier will conduct through R_2. The voltage drop caused by this error will appear as an input offset-voltage and noise source. But more critical, high input bias currents in the nano or microampere range may motivate you to lower your resistors in your circuit. When you do this, you will increase the capacitors in order to meet your filter cut-off frequency requirements. Large capacitors may not be a very good option because of cost, accuracy and size. Generally, filters with lower corner frequencies will require a CMOS amplifier instead of its bipolar amplifier counterpart.

Anti-Aliasing Filters for Near DC Analog Signals

Temperature measurement or flow measurement circuits belong to a class of applications where the analog signal is slow moving or nearly DC. In these types of sensing circuits, mechanical and environmental issues restrict the measurement speed. In temperature sensing circuits, the thermal time constant of the sensing elements (that is, thermistor, RTD or thermocouple) prevent the sensor from responding quickly to instantaneous temperature changes. Additionally, temperature seldom changes quickly. These circumstances keep most temperature sensing circuits operating near DC.

Chapter 4

RTD Temperature Sensor Circuit Revisited

Figure 4.17 shows an example of a precision temperature sensing circuit. In this circuit, a constant current source excites the sensing element, RTD. Chapter 3 covers the design and operation of this current source in detail. This circuit uses a single-supply, dual, CMOS amplifier. The amplifier closest to the RTD sensor (A1-1) eliminates wire-resistance, thermal errors. The second amplifier in this circuit (A1-2) implements gain as well as a 2nd order low-pass filter. The cut-off frequency of this filter should be as low as possible, and the transition region (the frequency region from the cut-off frequency to the frequency of the minimum desired attenuation) as small as possible. With this strategy, the analog filter eliminates most of the noise in the electronics.

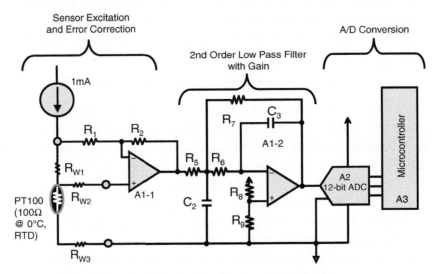

Figure 4.17: An RTD sensing circuit typically measures temperature at a low frequency rate. The anti-aliasing filter implemented in this circuit is 2nd order.

The full-scale range of the 12-bit, 100 ksps converter is V_{REF}. The 5 V power supply voltage connects to V_{REF}. For this system, a 10 ksps sampling rate is used. This sampling rate optimizes power dissipation of the converter. Table 4.4 summarizes the frequency response of Chebyshev, Butterworth and Bessel filters.

Table 4.4: Frequency response of five, low-pass filter designs which all have a $f_{CUT-OFF}$ = 1 Hz and A_{MAX} = 74 dB.

	FSTOP
Butterworth	71.5 Hz
Bessel	91.0 Hz
Chebyshev (−0.5 dB ripple)	85.5 Hz
Chebyshev (−2.0 dB ripple)	57.4 Hz
Chebyshev (−3.0 dB ripple)	50.6 Hz

The ideal signal-to-noise ratio of a 12-bit ADC is 74 dB. Although it may seem a little aggressive, I am going to make A_{MAX} equal to 74 dB for this design. The stop-band frequency (f_{STOP}) for the 2nd order filter should be 50 Hz or lower. This will eliminate 50 Hz and 60 Hz line noise. In order to keep the capacitor values reasonably low, the cut-off frequency is equal to 1 Hz.

The frequency response of the Butterworth and Chebyshev with a –0.5 dB. Both responses show 3rd order filters.

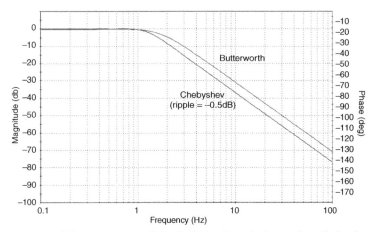

Figure 4.18: Of the Butterworth and Chebyshev designs, the Chebyshev exhibits the fastest roll-off over frequency after the cut-off frequency of the filter. Since this is a DC application, the Chebyshev is the best candidate.

It should be noticed in Figure 4.18 that the transition region (the frequency region after the cut-off frequency) of the Chebyshev filter has the most rapid attenuation over frequency. This type of response is most appropriate for reducing unwanted high frequency circuit noise in a precision application. The Chebyshev filter also allows the designer to design a lower order filter, therefore reducing circuit cost.

The anti-aliasing filter in this circuit is implemented with R5, R6, R7, R8, C2, C3 and the op amp A1-2. The gain stage, as well as a 2nd order low-pass filter is implemented with the following elements:

R1, R2 = 100 kΩ
R5 = 92.9 kΩ
R6 = 95.8 kΩ
R7 = 790 kΩ
R8 = 10 kΩ
R9 = 10 kΩ
C2 = 4.7 µF
C3 = 0.047 µF
Cut-off = 1 Hz, G = 24

Because the measurement is DC in nature, the phase distortions from the low-pass filter are negligible. Increasing the number of poles can further reduce the transition region from $f_{CUT-OFF}$ to f_{STOP}. This in turn increases the order of the filter.

Multiplexed Systems

At the other end of the spectrum, applications that have step functions in the analog signal path offer a different challenge to the low-pass, anti-aliasing filter. Figure 4.19 shows a common example of this type of application. In this circuit, the inputs of the multiplexer can have a combination of AC or DC signals. The most difficult moment of stabilization is immediately after a channel changes where the input to the filter appears as a step type function. An example of this type of situation would be a switch from CH0 to CH1, where the input to the filter changes from 1 V to 3 V or CH1 back to CH0. In this type of circuit, the most critical filter response characteristics are in the time domain. The anti-aliasing filter must be able to slew and settle quickly to a legitimate voltage. Otherwise, the converter must wait for a valid voltage input.

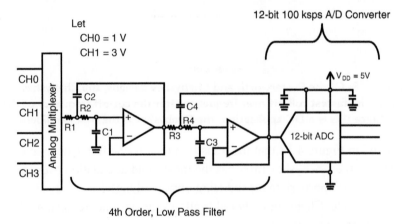

Figure 4.19: This system uses a multiplexer to configure four input signals into one ADC. When the multiplexer switches from one channel to another, it is possible that the 4th order anti-aliasing filter will be required to respond to a step function. An alternative to this solution is to move the low-pass filter to the other side of the multiplexer. Under this new configuration, the circuit would require four anti-aliasing filters instead of the one shown here.

Consider a two V-step response of a 4th order Butterworth, Bessel or Chebyshev filter (–0.5 dB ripple) in Figure 4.20. The corner frequency of all three of these filters is 1000 Hz.

In this example, the Chebyshev filter has a good frequency response, but also excessive ringing to a step response. The Butterworth filter is a better filter for the application, however, it does exhibit a fair amount of overshoot. The Bessel filter, step-response is controlled and quickly comes to a final value.

Do I Filter Now, Later or Never?

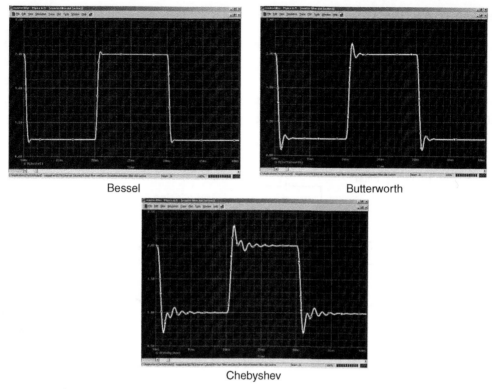

Figure 4.20: The time response of these three types of filters tell the tale when you are trying to decide which one to use in a multiplexed application circuit. I vote for the Bessel for the circuit in Figure 4.19.

Although the Bessel filter has the best step response of these three filters, the pass-band response in the frequency domain is not quite so good. Therefore, you will often find that you need a higher order filter when using Bessel designs.

From the diagrams in Figure 4.20, you may quickly conclude that a higher-order Bessel low-pass filter is best suited for the multiplexed circuit example.

You can implement this Bessel anti-aliasing filter with the following elements:

$R1 = 5.11$ kΩ
$R2 = 16.2$ kΩ
$R3 = 3.32$ kΩ
$R4 = 8.87$ kΩ
$C1 = 0.01$ µF
$C2 = 0.015$ µF
$C3 = 0.01$ µF
$C4 = 0.033$ µF

Chapter 4

Continuous Analog Signals

A third class of circuits needing an anti-aliasing filter is the case where the analog signal is continuous. Examples of this type of application would be a light sensing circuit or a system that is measuring the rpm of a motor. In these types of circuits, signal distortion is an issue from the point of view that you need to eliminate high frequency noise and still not contaminate the desired signal. Figure 4.21 shows an example of a light sensing circuit.

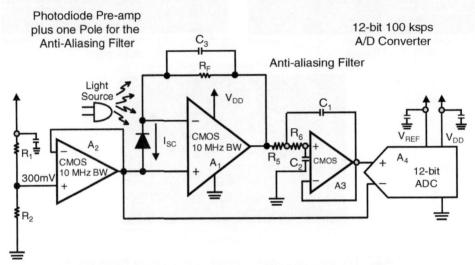

Figure 4.21: When light impinges on the photodiode, a charge is generated. This charge passes through the feedback resistor, R_F. The signal is then filtered with a 2nd order low-pass filter prior to the ADC.

With this circuit, the light that impinges on the photo detector generates charge or current over time. The photo detector's current flows through the feedback resistor, R_F, which generates a positive voltage at the output of the photo detector pre-amp. Chapter 3 contains details about the operation of this circuit. In this chapter, we discuss the filtering strategy.

The pre-amp not only amplifies the sensor signal, but it has a 1st order filter. This 1st order filter is implemented with the feedback capacitor, C_F. A 3rd order, low-pass filter follows the photo detector, pre-amplifier circuit. This filter should be flat over the lower frequencies in the pass band and then be able to provide an adequate roll-off or attenuation over frequency. Any reasonable overshoot or ringing in the time domain from the anti-aliasing filter is not noticeable because the photo detector, pre-amplifier circuit limits the input signal bandwidth. However, the sampling system will slow down if the filter has a long settling time. This middle-of-the-road application requires good attenuation in the frequency response, stop-band area of the filter, as well as minimal ringing in its time response.

The best type of filter for this circuit is the Butterworth. The Butterworth filter is the middle of the road filter that has average roll off in the frequency domain and minimal ringing in the time domain.

Matching the Anti-Aliasing Filter to the System

The RTD (Figure 4.17), multiplexed (Figure 4.19) and photo detector (Figure 4.21) applications suggest that the magnitude versus frequency response of a filter does not always give enough information. Going beyond the frequency behavior of filters, time domain issues such as group delay, overshoot, propagation delay and settling time have a significant affect on the overall filter performance, and consequently, the total data acquisition system.

The good news it that there are a variety of tools available to get you through the logistics (or painful calculations) of implementing discrete low-pass filters. Below is a list of free filter design programs from operational amplifier manufacturers that can be downloaded and used:

Active Filter Synthesis program	www.circuitsim.com
FilterPro™ program	www.ti.com
FilterLab® program	www.microchip.com
FilterCAD™ program	www.linear-tech.com
FilterWizard program	www.analog.com

But a word of warning concerning these filter programs; like a good tax preparation program, you will not maximize your refund potential unless you understand the fundamentals of tax law. And with filter design tools, the same holds true. It is easy to design the optimum filter circuit if you know the basics. (You will never guess what month of the year I wrote this chapter!)

So once you have a rough draft of your application circuit, you might ask, "What issues are important when designing the right anti-aliasing filter to drive my ADC?" SPICE simulations are used in this book to illustrate the intricacies of designing anti-alias filters.

Chapter 4 References

Analog Filter Design, Valkenburg, M. E. Van, Oxford University Press.

Active and Passive Analog Filter Design, An Introduction, Huelsman, Lawrence P., McGraw-Hill, Inc.

FilterLab, Analog filtering software tool at www.microchip.com.

"FilterPro™, MFB and Sallen-Key Low-Pass Filter Design Program," Bishop, Trump, Stitt, SBFA001A, Texas Instruments.

"Active Filters for Video meet Antialiasing and Reconstruction Requirements," Stutz, Bekgran, EDN, June 12, 2003.

"Switched-Capacitor Filters Beat Active Filters at Their Own Game," Yager, Laber, Techonline.

CHAPTER 5

Finding the Perfect Op Amp for Your Perfect Circuit

CHAPTER 5

Finding the Perfect Op Amp for Your Perfect Circuit

The operational amplifier's operation and circuits are easy to find in the books in your local university library. The amplifier operation and circuit descriptions found in these reference books take you through computational algorithms that theoretically will provide the solutions to your analog amplifier design woes. If there were a perfect amplifier on the market today, the designs found in these books would indeed be easy to implement successfully. But there isn't a perfect amplifier—yet. Throughout the history of analog system design, circuits have required special care in key areas in order to ensure success. As luck would have it, a little common sense and bench sense will pull you out of most of your amplifier, design disasters.

In an ideal world, the perfect amplifier would look like the one described in Figure 5.1. (See Appendix C for definitions of specifications.)

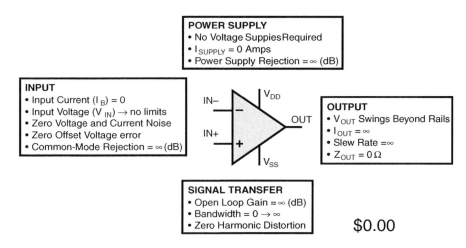

Figure 5.1: A perfect amplifier has an infinite input impedance, open-loop gain, power supply rejection ratio, common-mode rejection ratio, bandwidth, slew rate and output current. It also has zero offset voltage, input noise, output impedance, power dissipation and most importantly, zero cost.

The input stage design of this perfect amplifier would use devices whose inputs (IN+ and IN−) extend all the way to the power supply rails. Some single-supply amplifiers are able to do this with some distortion, but the perfect amplifier would be distortion-free. As a matter of fact, it would be nice if the inputs operated beyond the rails. If this were the case, the common-mode range goes beyond the rails as well.

Additionally, the inputs would not source or sink current—that is, they would have zero-input bias current. This allows source impedances to the amplifier to be infinite. This implies no common-mode or differential-mode input capacitance. Since voltage errors across the two inputs are usually gained by closed-loop circuit configurations around the amplifier, any DC voltage error (offset voltage) or AC error (noise) would be zero. The absence of these errors removes all of your calibration worries!

As for the power supply requirements of this ideal amplifier, there would be none. As you know, industry trends are always working on requests for lower supply voltages, and consequently, lower power consumption from active components. The ideal amplifier wouldn't need a voltage supply across V_{DD} and V_{SS} and would have zero power dissipation in its quiescent state.

The output of this amplifier would be capable of really swinging rail-to-rail, or even beyond. This would eliminate the problem of losing bits on the outer rim in the following A/D conversion. The output impedance would be zero at DC, as well as over frequency ensuring that the device connected to the input of the amplifier is perfectly isolated from the external output device. The op amp would respond to input signals instantaneously—that is, the slew rate would be infinite and it would be able to drive any load (resistive or capacitive) while maintaining an infinite open-loop gain and rail-to-rail output swing performance. Finally, in the frequency domain, the open-loop gain would be infinite at DC as well as over frequency, and the bandwidth of the amplifier would also be infinite. Oh, did I forget price? We would all love to have this ideal amplifier for $0.00.

Welcome to OP AMP 101! This describes the textbook amplifier.

I know that if I'm able to figure out how to design this amplifier, I guarantee you, I will become a multizillionaire. At this point, you are probably saying "only in your dreams!" Well, maybe not a multizillionaire, mainly because the profits are $0.00. However, it is certain that I will become a very popular (still poor) person.

It is interesting to note that many of these design imperfections are used to an advantage by most designers. For example, an amplifier circuit design uses a less than infinite bandwidth to limit the noise and high-speed transients in circuits. An infinite slew rate is not as good as it sounds. The amplifier users enjoy slower signals. This reduces the glitches further on down the signal path and simplifies your layout.

So, for today, we know that there isn't an ideal amplifier for all circuit situations. The best we can do with the choices available is to pick the best amplifier for our application circuit and then use it properly.

Finding the Perfect Op Amp for Your Perfect Circuit

Choose the Technology Wisely

CMOS and bipolar are the two silicon technologies that single-supply operational amplifiers commonly use. Figure 5.2 shows the differences between these two operational amplifier technologies. The most important difference between CMOS and bipolar is in the input stage transistors. These transistors have a profound effect on the overall operation of the amplifier.

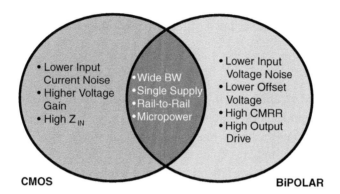

Figure 5.2: The two silicon technologies that single-supply amplifiers are manufactured with are CMOS or bipolar processes. By using the CMOS process, you can manufacture bipolar amplifiers. In these designs, the input transistors are bipolar, and the remaining transistors are CMOS.

Because of the difference between the input transistors of these two types of amplifiers, the CMOS amplifier has lower input current noise and higher input impedance. Because of the high input impedance, the input bias current of the CMOS amplifier is much lower. In fact, the electrostatic discharge (ESD) cells at the input of the CMOS amplifier causes the input bias current errors. As will be shown in circuits later in this chapter, we can use this to an advantage for high impedance sources, such as photosensing transimpedance amplifiers.

The CMOS amplifier typically has a higher open-loop gain than bipolar amplifiers. This can minimize gain error in applications where the closed-loop gain is extremely high (60 dB or greater).

In contrast with the CMOS amplifier, the bipolar amplifier usually has lower input-voltage noise, room temperature offset-voltage and offset-drift. Bipolar amplifiers are more likely to provide higher output drive. They also exhibit a higher common-mode rejection capability. This is useful if the amplifier is in a buffer configuration. Although these specifications are typically better than the CMOS amplifier counterpart, the input bias current and input current noise is considerably higher.

Single-supply operating conditions are perfect for both CMOS and bipolar amplifiers. With the proper IC design, they are also capable of input and output rail-to-rail operation.

Chapter 5

Fundamental Operational Amplifier Circuits

The op amp is the analog building block that is analogous to the digital gate. By using the op amp in the design, circuits can be configured to modify the signal in the same fundamental way that the inverter, AND and OR gates do in digital circuits. This section of the chapter will show the fundamental circuits using this building block. The list of circuits we will discuss include the voltage follower, noninverting gain and inverting gain circuits. This will be followed by more complex circuits, including a difference amplifier, summing-amplifier and current-to-voltage converter.

Voltage Follower Amplifier

Starting with the most basic op amp circuit, the buffer amplifier (shown in Figure 5.3) is used to drive heavy loads, solve impedance matching problems, or isolate high power circuits from sensitive, precise circuitry. Usually, heavy loads require an additional specialized amplifier that is capable of supplying the higher output currents that are greater than 20 mA. You will find that the amplifier data sheet has specifications for the magnitude of the amplifier output current, capable of driving higher currents.

Solving impedance matching problems is also a good reason to use a buffer amplifier. This type of problem exists when the signal path has a high impedance device or resistor that creates an undesirable voltage divider in the circuit. A buffer amplifier breaks up this type of impedance path because of the high impedance input and low impedance output of the amplifier.

Another use for a buffer is to keep high thermal changes away from sensitive circuits. In this scenario the buffer follows the sensitive circuit and serves the purpose of driving high output currents.

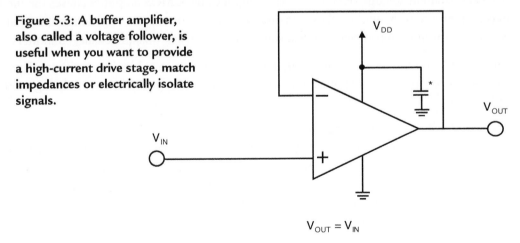

Figure 5.3: A buffer amplifier, also called a voltage follower, is useful when you want to provide a high-current drive stage, match impedances or electrically isolate signals.

$V_{OUT} = V_{IN}$

* Bypass capacitor, 1 µF or 0.1 µF

The buffer amplifier as shown in Figure 5.3 can be implemented with any single-supply, unity gain stable amplifier. In this circuit, as with all amplifier circuits, bypassing the op amp power with a capacitor is a must. For single-supply amplifiers that operate in bandwidths from DC to 1 MHz, a 1 µF capacitor is usually appropriate. Sometimes a smaller bypass capacitor is required for amplifiers that have bandwidths of up to the tens to hundreds of megahertz. In these cases, a 0.1 µF capacitor would be appropriate. If the selection of the value of the bypass capacitor is improper, the op amp may oscillate.

The analog gain of the circuit in Figure 5.3 is +1 V/V. Notice that this circuit has positive overall gain, but the feedback loop is tied from the output of the amplifier to the inverting input. An all too common error is to assume that an op amp circuit that has a positive gain requires positive feedback. You can configure this amplifier with positive feedback if you connect the noninverting input to the output. I know this sounds unbelievable, but I have had applicants draw buffers with positive feedback during interview. If positive feedback is used, the amplifier will most likely drive to either rail at the output.

This amplifier circuit will give good linear performance across the bandwidth of the amplifier. And, you may be looking at this discussion and saying to yourself, "There is that textbook description, again." You are right, however, here are the land mines in this type of circuit.

The only restrictions on the signal will occur as a result of a violation of the input common-mode voltage and output swing limits. You need to scrutinize these performance characteristics in your amplifier data sheet and your application's demands on this type of circuit. Oh, by the way, ensure that the bandwidth of the amplifier is at least 100× higher than the bandwidth of your signal. However, be aware that you need to look at the input and output of the amplifier. Chapter 6 discusses these limitations in detail.

When using this circuit to drive heavy loads, the specifications of the amplifier must indicate that it is capable of providing the required output currents. Another application where this circuit may be used is to drive capacitive loads. Not every amplifier is capable of driving capacitors without becoming unstable. If an amplifier can drive capacitive loads, the product data sheet will highlight this feature. However, if an amplifier can't drive capacitive loads, the product data sheets will not explicitly say so. This is an instance where features are not in the advertisements or promotions and there is no mention of average performance.

Another use for the buffer amplifier is to solve impedance-matching problems. This would be applicable in a circuit where the analog signal source has relatively high impedance as compared to the impedance of the following circuitry. If this occurs, there will be a voltage loss with the signal because of the voltage divider between the source's impedance and the following circuitry's impedance. The buffer amplifier is a perfect solution to the problem. The input impedance of the noninverting input of an amplifier can be as high as 10^{13} Ω for CMOS amplifiers. In addition, the output impedance of this amplifier configuration is usually less than 100 Ω.

Chapter 5

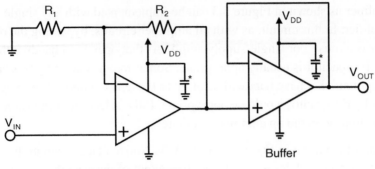

*Bypass capacitor, 1 µF or 0.1 µF

Figure 5.4: A buffer amplifier helps achieve load isolation in this circuit. The buffer separates any high-current output requirements from this input amplifier.

Yet, another use of this configuration is to separate a heat source from sensitive precision circuitry, as shown in Figure 5.4. Imagine that the input circuitry to this buffer amplifier is amplifying a 100 mV signal. This type of amplification is difficult to do with any level of accuracy in the best of situations. Assigning the output current drive to the device that is doing the precision, amplification work can easily disrupt this measurement. An increase in current drive will cause self-heating of the chip, which will induce an offset change. In this circuit (Figure 5.4), the front-end circuitry makes precision measurements, while an analog buffer performs the function of driving a heavy load.

Gaining Analog Signals

The buffer solves many analog signal problems; however, there are instances in circuits where you need to gain a signal. Two fundamental types of amplifier circuits can provide gain. With the first type, the signal gain is positive (or not inverted) as shown in Figure 5.5. This type of circuit is useful in single-supply amplifier applications where negative voltages are usually not present, difficult to produce or just not possible.

The input signal to this circuit is presented to the high impedance, noninverting input of the op amp. The gain that the amplifier circuit applies to the signal is equal to:

$$V_{OUT} = (1 + R_1/R_2) V_{IN}$$

Typical values for these resistors in single-supply circuits are above 5 kΩ to 25 kΩ for R_2. The input resistor, R_1, restrictions are dependent on the amount of gain desired versus the amount of amplifier noise and input offset voltage as specified in the product data sheet of the op amp.

Again, this circuit has some restrictions in terms of the input and output range. The common-mode range of the amplifier restricts the noninverting input. The output swing of the amplifier is also restricted as stated in the product data sheet of the individual amplifier. Most typically,

Finding the Perfect Op Amp for Your Perfect Circuit

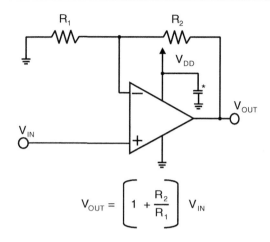

Figure 5.5: This is an operational amplifier configured in a noninverting gain circuit. This circuit applies a positive gain to a signal in your circuit. Therefore, you won't need a reference level-shift voltage to keep the output of the amplifier within its operating range.

$$V_{OUT} = \left[1 + \frac{R_2}{R_1} \right] V_{IN}$$

* Bypass capacitor, 1 µF or 0.1 µF

the larger signal at the output of the amplifier causes more signal-clipping errors than the smaller signal at the input. Reducing the gain of this circuit may eliminate undesirable output clipping errors.

Figure 5.6 illustrates an inverting amplifier configuration. This circuit gains and inverts the signal present at the input resistor, R_1. The gain equation for this circuit is:

$$V_{OUT} = -(R_2 / R_1) V_{IN}$$

The ranges for R_1 and R_2 are the same as in the noninverting circuit shown in Figure 5.5.

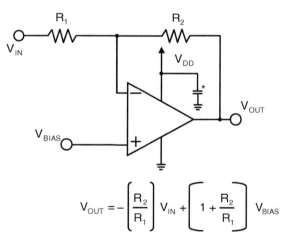

Figure 5.6: This is an operational amplifier configured in an inverting gain circuit. Single-supply environments require V_{BIAS} to ensure the output stays above ground.

$$V_{OUT} = -\left[\frac{R_2}{R_1} \right] V_{IN} + \left[1 + \frac{R_2}{R_1} \right] V_{BIAS}$$

* Bypass capacitor, 1 µF or 0.1 µF

Chapter 5

This circuit has a minor pitfall in single-supply circuits. In single-supply applications, this circuit is easy to misuse. The problem is rooted in the selection of the voltage at V_{BIAS}. You need to select a value for V_{BIAS} so that the output of the amplifier always remains between the supplies.

For example, let R_2 equal 10 kΩ, R_1 equal 1 kΩ, V_{BIAS} equal 0 V, and the voltage at the input resistor, R_1, equal to 100 mV, the output voltage would be –1 V. This would violate the output swing range of the operational amplifier. In reality, the output of the amplifier would try to go as near to ground as possible.

The inclusion of a positive DC voltage at V_{BIAS} in this circuit solves this problem. In the previous example, a voltage of 225 mV applied to V_{BIAS} would level shift the output signal up 2.475 V. This would make the output signal equal (2.475 V – 1 V) or 1.475 V at the output of the amplifier. Typically, you want to make the target average output voltage of the amplifier equal to $V_{DD}/2$.

The Difference Amplifier

The difference amplifier combines the noninverting amplifier and inverting amplifier circuits of Figure 5.5 and Figure 5.6 into a signal block that subtracts two signals. Figure 5.7 illustrates an example of the difference amplifier circuit.

Figure 5.7 illustrates a straightforward implementation of this function. A difference amplifier or op amp subtractor uses this arrangement of resistors around an amplifier. The DC transfer function of this circuit is equal to:

$$V_{OUT} = V_{IN+} \, R_4(R_1 + R_2) / ((R_3 + R_4)R_1) - V_{IN-}(R_2/R_1) + V_{SHIFT}$$
$$R_3(R_1 + R_2)/((R_3 + R_4)R_1)$$

If R_1/R_2 is equal to R_3/R_4, the closed loop system gain of this circuit equals:

$$V_{OUT} = (V_{IN+} - V_{IN-})(R_2/R_1) + V_{SHIFT}$$

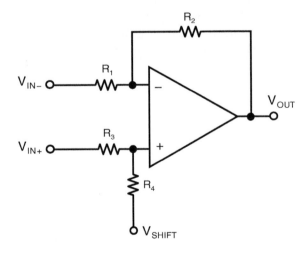

Figure 5.7: This is an operational amplifier circuit configured in a difference amplifier circuit. A difference amplifier implements the subtraction and gain function in one stage.

Finding the Perfect Op Amp for Your Perfect Circuit

This circuit configuration will reliably take the difference of two signals as long as the signal-source impedances are low. If the signal source impedances are high with respect to R_1, there will be a signal loss due to the voltage divider action between the source and the input resistors to the difference amplifier. Additionally, errors can occur if the two signal source impedances are mismatched. With this circuit, it is possible to have gains equal to or higher than one.

The fact that R_1/R_2 is equal to R_3/R_4 simplifies the mathematics in this system considerably. Since the gain of both signals is equal, the difference amplifier conveniently subtracts the common-mode voltage of the two signals from the system. It is also easy to implement gain by setting the two resistor ratios to be equal or greater than one.

One limitation of this circuit is the lack of flexibility with gain adjustments. If you change the gain dynamically in the application, you must adjust two resistors. In a single-supply environment, a voltage reference centers the output signal between ground and the power supply. Figure 5.7 shows this voltage, "V_{SHIFT}". The purpose of this reference voltage is to simply shift the output signal into the linear region of the amplifier. A precision, voltage-reference, or a resistive network implements the V_{SHIFT} circuit function as shown in Figure 5.8.

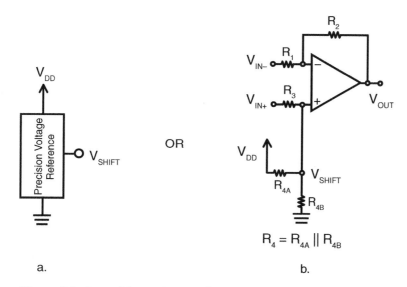

Figure 5.8: A precision voltage reference, (a) or a less expensive solution of replacing R_4 of the voltage divider between the supply, (b) provides the voltage, V_{SHIFT}, of this difference amplifier.

Summing Amplifier

You can use summing amplifiers to combine multiple signals by addition or subtraction. Since the difference amplifier can only process two signals, it is a subset of the summing amplifier.

Figure 5.9: Operational amplifier configured in a summing amplifier circuit.

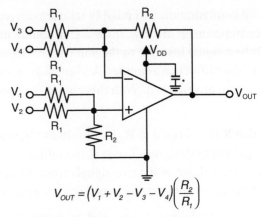

$$V_{OUT} = (V_1 + V_2 - V_3 - V_4)\left(\frac{R_2}{R_1}\right)$$

* Bypass capacitor, 1 µF or 0.1 µF

The transfer function of this circuit as shown in Figure 5.9 is:

$$V_{OUT} = (V_1 + V_2 - V_3 - V_4)(R_2 / R_1)$$

You can use any number of inputs on either the inverting or noninverting input sides as long as there are an equal number of both with equivalent resistors.

Current-to-Voltage Conversion

If you use a photodetector, feedback resistor and an operational amplifier in your circuit you can sense light. This type of circuit converts the output current of a photodetector into a voltage. The single resistor and an optional capacitor are in the feedback-loop of the amplifier as shown in Figure 5.10.

Figure 5.10: These circuits show how to convert current to voltage by using an amplifier and one resistor. The top light-sensing circuit is appropriate for precision applications. The bottom circuit is appropriate for high-speed applications.

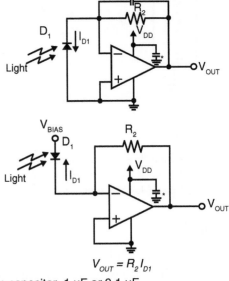

$$V_{OUT} = R_2 I_{D1}$$

* Bypass capacitor, 1 µF or 0.1 µF

In the circuits shown in Figure 5.10, light impinging on the photodetector generates a current. This current flows in the reverse bias direction of the diode. If a CMOS op amp is used, the high input impedance of the op amp causes the current from the detector (I_{D1}) to go through the feedback resistor, R_2. Additionally, the op amp input bias current error is low because it is CMOS (typically <200 pA). You would ground the noninverting input of the op amp, which keeps the entire circuit biased to ground. These two circuits will only work if the common-mode range of the amplifier includes zero and you are not concerned about a zero level of light. If your light source has zero luminance, the output of the single-supply amplifier is unable to go all the way to ground. The circuit in Chapter 3, Figure 3.13 illustrates a solution to this problem.

The two circuits in Figure 5.10 provide precision-sensing from the photodetector (top circuit in the figure) and higher speed sensing (bottom circuit in the figure). In the top circuit, the voltage across the detector is nearly zero and equal to the offset voltage of the amplifier. With this configuration, current that appears across the resistor, R_2, is primarily a result of the light excitation on the photodetector.

The photosensing circuit at the bottom of the figure works best in a high-speed, digital environment. By reverse biasing the photodetector (which reduces the parasitic capacitance of the diode), this sensing circuit can respond very quickly to digital signals. There is more leakage through the photodetector in this bottom circuit, which causes a higher DC error.

Using these Fundamentals

Instrumentation Amplifier

You will find instrumentation amplifiers in a large variety of applications from medical instrumentation to process control. The instrumentation amplifier is similar to the difference amplifier in that it subtracts one analog signal from another, but it differs in terms of the quality of the input stage. Figure 5.11 illustrates a classic, three op-amp instrumentation amplifier.

Figure 5.11: You can design an instrumentation amplifier with three amplifiers. The input operational amplifiers (A_1, A_2) provide signal gain. The output operational amplifier converts the signal from the two input amplifiers to a single-ended output with a difference amplifier (A_3).

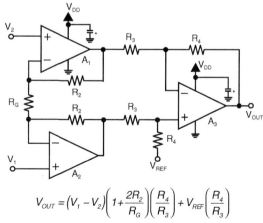

$$V_{OUT} = (V_1 - V_2)\left(1 + \frac{2R_2}{R_G}\right)\left(\frac{R_4}{R_3}\right) + V_{REF}\left(\frac{R_4}{R_3}\right)$$

* Bypass capacitor, 1 µF or 0.1 µF

Chapter 5

In this circuit, the high impedance, noninverting inputs of the input amplifiers (A_1, A_2) acquire the two input signals. This is a distinct advantage over the difference amplifier configuration where source impedances are high or mismatched. The first stage also gains the two incoming signals. One resistor, R_G, adjusts the gain.

Following the first stage of this circuit is a difference amplifier (A_3). The function of this portion of the circuit is to reject the common-mode voltage of the two input signals as well as differentiate them. The source impedances of the signals into the input of the difference amplifier are low, equivalent and well controlled.

The reference voltage (V_{REF}) of the difference stage of this instrumentation amplifier is capable of spanning a wide range. Typically, you would connect the voltage reference to half of the supply voltage in a single-supply application. The transfer function of this circuit is:

$$V_{OUT} = (V_1 - V_2)(1 + 2R_2/R_G)(R_4/R_3) + V_{REF}(R_4/R_3)$$

Figure 5.12 shows a second type of instrumentation amplifier. In this circuit, the two amplifiers serve the functions of load isolation and signal gain. The second amplifier also differentiates the two input signals (V_1, V_2).

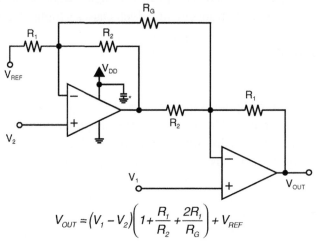

$$V_{OUT} = (V_1 - V_2)\left(1 + \frac{R_1}{R_2} + \frac{2R_1}{R_G}\right) + V_{REF}$$

* Bypass capacitor, 1 µF or 0.1 µF

Figure 5.12: You can design an instrumentation amplifier with two amplifiers. This configuration is best suited for higher gains (gain ≥ 3 V/V).

You would connect the circuit reference voltage to the first op amp in the signal chain. Typically, this voltage is half of the supply voltage in a single-supply environment.

The transfer function of this circuit is:

$$V_{OUT} = (V_1 - V_2)(1 + R_1/R_2 + 2R_1/R_G) + V_{REF}$$

Floating Current Source

A floating current source can come in handy when driving a variable resistance, like a resistance temperature device (RTD). This particular configuration produces an appropriate 1 mA source for an RTD-type sensor. However, you can change this current reference magnitude to any current.

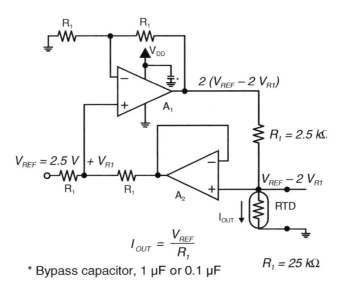

Figure 5.13: A floating current source uses two operational amplifiers and a precision voltage reference.

With this configuration, R_1 reduces the voltage of V_{REF} by the voltage V_{R1}. The voltage applied to the non-inverting input of the top op amp is $V_{REF} - V_{R1}$. This voltage is gained to the amplifier's output by two, to equal $2 \times (V_{REF} - V_{R1})$. Meanwhile, the output for the bottom op amp (A_2) is presented with the voltage $V_{REF} - 2V_{R1}$. Subtracting the voltage at the output of the top amplifier from the non-inverting input of the bottom amplifier gives $2 \times (V_{REF} - V_{R1}) - (V_{REF} - 2V_{R1})$, which equals V_{REF}.

The transfer function of the circuit is:

$$I_{OUT} = V_{REF} / R_I$$

Amplifier Design Pitfalls

Theoretically, the circuits within this chapter work. Beyond the theory, however, there are few tips that will help get the circuit right the first time. In this section, along with the discussion in Chapter 6, lists common problems associated with using an op amp on a PC board. The following discussion has two categories: general suggestions and single-supply pitfalls.

Chapter 5

In General

- Be careful of the supply pins. Don't make them too high per the amplifier specification sheet and don't make them too low. High supplies will damage the part. In contrast, low supplies won't bias the internal transistors and the amplifier won't work or it may not operate properly.

- Make sure the negative supply (usually ground) is, in fact, tied to a low impedance potential. Additionally, make sure the positive supply is the voltage you expect with respect to the negative supply pin of the op amp. Placing a voltmeter across the negative and positive supply pins will verify that you have the right relationship between the pins.

- Ground can't be trusted, especially in digital circuits. Plan your grounding scheme carefully. If the circuit has a lot of digital circuitry, consider separate ground and power planes. It is very difficult, if not impossible, to remove digital switching noise from an analog signal.

- Bypass the amplifier power supplies with bypass capacitors as close to the amplifier as possible. CMOS amplifiers usually use a 1 µF or 0.1 µF capacitor. Also, bypass the power supply with a 10 µF capacitor.

- Use short lead lengths to the inputs of the amplifier. If you have a tendency to use the white perf boards for prototyping, be aware that they can cause noise and oscillation. There is a good chance that these problems won't be a problem with the PCB implementation of the circuit.

- Amplifiers are static sensitive! If damage has occurred, they may fail immediately or exhibit a soft error (like offset voltage or input bias current changes) that will get worse over time.

Single-Supply Rail-to-Rail Amplifiers

- Operational amplifier output drivers are capable of driving a limited amount of current to the load. Check your product data sheet for that number

- Capacitive loading an amplifier is risky business. Make sure the amplifier can handle any loads that you may have.

- It is very rare that a single-supply amplifier will truly swing rail-to-rail. In reality, the output of most of these amplifiers can only come within 50 to 300 mV from each rail. Check the product data sheets of your amplifier.

Moving forward, the next level of troubleshooting amplifier tips is in the frequency domain. You will find with stability issues such attributes as ringing or oscillation. If you are interested in these issues, turn the page. Chapter 6 will provide some insight.

Chapter 5 References

Design with Operational Amplifiers and Analog Integrated Circuits, Sergio Franco, McGraw Hill.

Intuitive Operational Amplifiers, Frederiksen, Thomas, McGraw Hill.

Analog Circuit Design, Williams, Jim, Butterworth-Heinemann.

"Operational Amplifiers: 6 Part," Baker, Bonnie C., First Published in *analogZone* (2002, 2003) and reproduced with permission.

CHAPTER 6

Putting the Amp Into a Linear System

CHAPTER 6

Putting the Amp Into a Linear System

Your theories and textbook examples can only take you so far. It is useful to come to terms with the fact that amplifiers have imperfections. One of my mentors once said every amplifier is waiting to oscillate, and every oscillator is waiting to amplify. He is right, especially if you are working "blind" in terms of knowing the amplifier characteristics. This chapter will go into detail concerning AC and DC performance specifications in terms of the hidden parasitics, the theoretic laws that you should not break, and a few practical guidelines that ensure success the first time.

The first section of this chapter discusses what the inputs and output of the single-supply amplifier does in a static, DC environment. We will start with the input stage and talk about the transistor topology. Each topology we see will give you a hint about the overall performance of the input stage of the amplifier. Following the input stage discussion, we will touch on output stage characteristics. Once again, you need to evaluate your output stage at very low frequencies so that the bandwidth of the amplifier is not affecting the performance of the amplifier. And finally, there will be a short summary about specifications such as the offset voltage, open-loop gain, common-mode rejection ratio, and power supply rejection ratio. The second half of this chapter will cover the AC characteristics of the amplifier.

The Basics of Amplifier DC Operation

Amplifier Input Stage Anomalies

The transmission of a signal through the operational amplifier begins at the input stage. When you select your amplifier, you should first scrutinize the characteristics of the external input signal. For example, what voltage range would you expect your source signal to span? If the voltage range of the signal spans from one power supply rail to the other and you can easily tolerate a positive gain, a rail-to-rail input amplifier would be appropriate for your application. If not, you can probably settle for an amplifier that does not have a rail-to-rail input operation. You should know, however, if you are using a rail-to-rail input amplifier, that an offset voltage distortion occurs as you take the input across its entire common-mode range. Figure 6.1 shows an example of the distortion of a rail-to-rail input CMOS amplifier.

Chapter 6

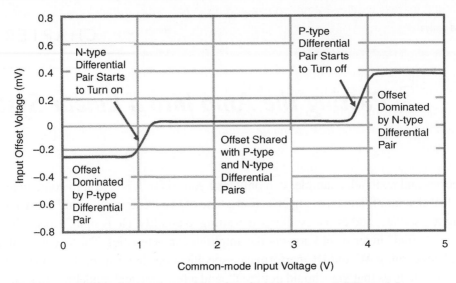

Figure 6.1: Rail-to-rail input amplifiers require two pairs of differential transistors (composite input stage, Figure 6.2c). Consequently, they have an input offset distortion as the input common-mode voltage passes through the regions where one or the other pair is turning on or off. Although the amplifier represented with this graph has two offset voltage transition regions, some single-supply, rail-to-rail input amplifiers only have one transition region.

The input transistor configuration of the rail-to-rail amplifier input causes this distortion. The input-voltage range is more a function of the input circuit topology rather than the silicon process. The input devices of the amplifier can be CMOS or bipolar. There are three basic CMOS input pair topologies: p-channel metal-oxide semiconductor (PMOS), n-channel metal-oxide semiconductor (NMOS), and PMOS/NMOS composite. The diagram in Figure 6.2 illustrates CMOS input topologies. Figure 6.2 also shows the basic topologies for bipolar-input amplifiers if you replace the NMOS and PMOS devices with NPN and PNP transistors, respectively. With bipolar-input stages, input-offset voltage variations are still a problem. However, the bipolar transistors introduce an additional input bias current error. The nano-ampere base current of an NPN (n-type collector, p-type base, n-type emitter) transistor goes into the device, while the nano-ampere base current of a PNP (p-type emitter, n-type base, p-type collector) transistor comes out of the device.

Figure 6.2a uses PMOS transistors (Q_1 and Q_2) for the first device at the input terminals. With this particular topology, the gate of both transistors can go 0.2 to 0.3 V below the negative power-supply voltage before these devices leave their active region. However, the input terminal limit is several hundred millivolts below the positive power-supply voltage as the input devices leave their normal region of operation. An amplifier designed with a PMOS input stage will typically have an input range of $V_{SS} - 0.2$ V to $V_{DD} - 1.2$ V.

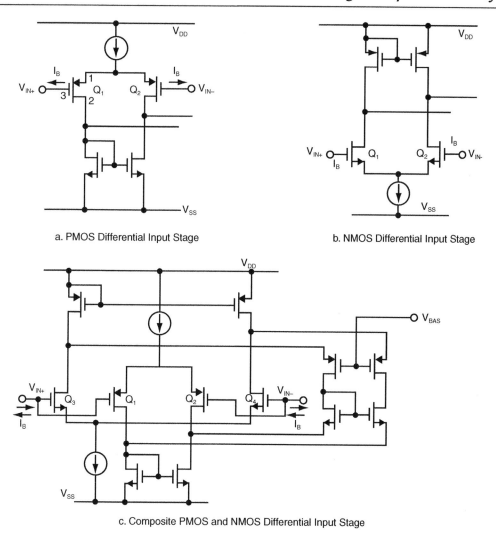

Figure 6.2: The input-voltage range of an amplifier is dependent on the topology of the input stage of the amplifier. The input stage can be constructed of PMOS, (a) differential pair allowing the input to swing below the negative supply or a NMOS differential pair, (b) where the inputs can swing above the positive supply. A composite input-stage, (c) uses PMOS and NMOS differential-pairs so the input-voltage range can extend above the positive rail to below the negative rail.

As shown in Figure 6.2b, the input range is restricted near the negative power-supply voltage if NMOS transistors are used. In this case, the input terminals range up to a few tenths of a volt above the positive-supply rail, but only to 1.2 V above the negative-supply rail. This single-supply amplifier topology is not as common as the topology in Figure 6.2a. This is because most signals in the analog world use ground as a reference point; not the positive supply.

Chapter 6

The limitations of the circuits in Figure 6.2a and 6.2b are usually not a problem, although you may think that you absolutely have to have a rail-to-rail input device. Most amplifiers in applications have a closed-loop gain of two or higher. As a result, the output of the amplifier will reach the positive rail sooner that the input will.

The most common circuits that require rail-to-rail amplifiers include buffer or instrumentation amplifier circuits. The configuration of a composite stage uses PMOS and NMOS transistors (Figure 6.2c). With this topology, the amplifier effectively combines the advantages of the PMOS and NMOS transistors for true rail-to-rail input operation. The PMOS transistors are turned completely on, and the NMOS transistors are completely off when the input terminals of the amplifier are driven towards the negative rail. Conversely, the NMOS transistors are in use while the PMOS transistors are off when driving input terminals are to the positive rail.

Although, this style of input stage has rail-to-rail input operation, there are performance compromises. This design topology will have wide variations in offset voltage. In the region near ground, the offset error of the PMOS portion of the input stage is dominant. In the region near the positive power supply, the NMOS transistor pair dominates the input stage offset error.

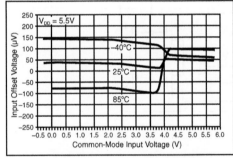

a. 10 MHz, CMOS, precision op amp
(MCP6021, Microchip Technologies)

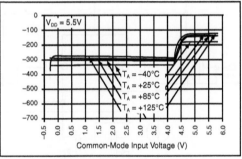

b. 1 MHz, low cost, CMOS op amp
(MCP6001, Microchip Technologies)

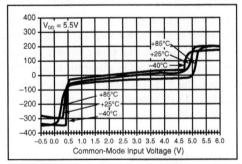

c. 14 kHz, ultra low power, CMOS op amp
(MCP6041, Microchip Technologies)

Figure 6.3: This graph shows the input-offset voltage versus common-mode voltage of three different CMOS operational amplifiers over temperature with a single-supply voltage of 5.5 V. All three amplifiers have rail-to-rail input swing capability.

The graphs in Figure 6.3 illustrate this performance characteristic nicely. With this topology, the offset voltage error can change dramatically in magnitude and sign, as the common-mode voltage of the amplifier inputs extend over their entire range.

Figure 6.3 shows the performance of three different CMOS amplifiers. The input stage topology for all three of these amplifiers is a composite PMOS and PMOS differential input stage, as illustrated in Figure 6.2c. The amplifier in Figure 6.3a is a high-speed, precision 10 MHz device. At lower input voltages, the PMOS portion of the input stage is in operation with the NMOS portion turned off. At approximately 4.0 V, the NMOS portion of the input stage starts to turn on and take over the operation of the input stage. The amplifier in Figure 6.3b is a 1 MHz amplifier. This device has the same input stage topology, but the NMOS portion of the input stage starts to take over at a slightly higher voltage, 4.4 V. Finally, the amplifier in Figure 6.3c has two transitions that occur as the common-mode input voltage is increased. At very low common-mode voltages, the PMOS portion of the input stage is operational. This also turns off the NMOS portion. As the common-mode input voltage increases, the NMOS section of the circuit quickly turns on. From approximately 0.5 V to 4.6 V, the PMOS and NMOS sections are operating. When the input voltage reaches approximately 4.75 V, the PMOS begins to shut down, leaving the NMOS stage operational.

If this offset distortion feature is not desirable, you may want to consider designing your amplifier circuit in an inverting gain configuration. Figure 6.4b shows an example of this type

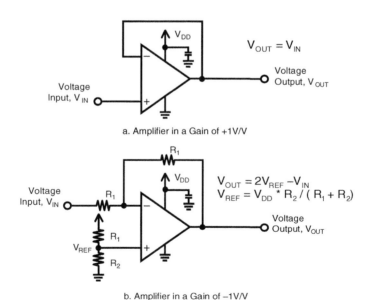

a. Amplifier in a Gain of +1V/V

b. Amplifier in a Gain of –1V/V

Figure 6.4: If the input-voltage swings rail-to-rail, an amplifier configured as a single-supply buffer or voltage follower (a) should have rail-to-rail input capability. However, if an amplifier circuit gain is negative (b) the common-mode voltage of the amplifier input will remain at V_{REF}.

Chapter 6

of circuit. The circuit shown in Figure 6.4b. is best suited for single-supply applications. This configuration produces a negative gain and adds a positive voltage to the mix. You can find the transfer function equation in the figure. In this circuit, the common-mode voltage of the amplifier is kept constant at $V_{DD} * R_2 / (R_1 + R_2)$. Therefore, you will not need an amplifier with rail-to-rail input capability. The amplifier common-mode range only needs to include $V_{DD} * R_2 / (R_1 + R_2)$.

However, don't let this offset distortion scare you away if you really need rail-to-rail inputs. With single-supply circuits, rail-to-rail input amplifiers are needed when a buffer amplifier circuit (Figure 6.4a) is used or possibly with an instrumentation amplifier configuration. Just be aware that if either of the inputs of the amplifier goes beyond the specified input range of that amplifier, the output will typically go to one of the power supply rails. No guarantee which rail.

High-Input Impedance May Make a Difference

An active filter uses a combination of one amplifier, one to three resistors, and one to two capacitors to implement one or two poles. The active filter offers the advantage of providing "isolation" between stages. This is possible by taking advantage of the high-input impedance and low-output impedance of the operational amplifier. In all cases, the number of capacitors at the input and in the feedback loop of the amplifier determines the order of the filter.

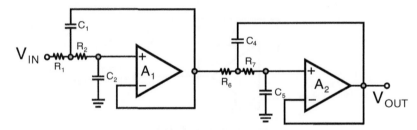

4th order (or 4-pole), low pass filter
using two, 2-pole, Sallen-Key low pass filters

Figure 6.5: If the corner frequency of a low-pass filter is low, you will need either high value resistors or high value capacitors. In this case, a bipolar amplifier (input bias current ~100s of nano-amps) will generate a voltage error as the input bias current flows through the input resistors ($[R_1 + R_1]$ and $[R_4 + R_7]$). Firmware or software solutions can correct this type of error. However, an easier solution is to use the right amplifier for the job. My choice would be the CMOS input amplifier, where the input bias currents at room temperature are typically less than 1 pA at room temperature.

You can implement this low-pass filter function by using resistor and capacitor pairs around the amplifier (Figure 6.5). The input bias current of the amplifiers, A_1 and A_2, flows through $(R_1 + R_1)$ and $(R_4 + R_7)$, inclusive. Bipolar amplifiers with PNP or NPN input differential

Putting the Amp Into a Linear System

pairs, will generate a bias current with a magnitude in the hundreds of nano-amps. A low-frequency, low-pass filter (1 Hz to 100 Hz corner frequencies, see Chapter 4 for more details about filters) requires either high-value capacitors or high-value resistors in the implementation. High-value capacitors are bulky and expensive. On the other hand, high-value resistors are not bulky and they cost the same as the lower-value resistors. It is not unusual for a 10 Hz (corner frequency) low-pass filter to have resistors as high as 500 kΩ to 1 MΩ. You can imagine that the error voltage generated by these resistors with bipolar amplifiers can be high. For example, if the input bias current of A_1 is 200 nA and ($R_1 + R_2$) equals 1.5 MΩ, the introduced offset error is 300 mV. If you change A_1 from a bipolar to CMOS amplifier with an input bias current of 1 pA, the introduced error is 1.5 µV. It is true that you can use your microcontroller to calibrate out the bipolar amplifier errors. But, why not pick the right amplifier for the job?

Hey, the Output Doesn't Swing Rail-to-Rail (as Promised)

Single-supply amplifiers do not truly swing rail-to-rail at the output. To make matters worse, at the outer regions (near the rail), the amplifier will behave in a nonlinear fashion. The reality of this performance characteristic is that the output of single-supply amplifiers can only come within 50 to 300 mV from each rail. Figure 6.6 illustrates this behavior.

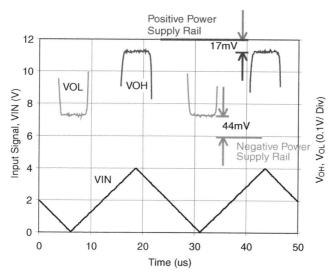

Figure 6.6: The bottom plot in this figure illustrates the input voltage swing to an amplifier that is configured in a gain of +2 V/V. The top plot shows the magnified output voltage of the amplifier. Included in the top portion of this plot is an indication where the positive supply rail is, with respect to V_{OH} and the negative supply rail with respect to V_{OL}.

Chapter 6

The advertisements that claim of a "rail-to-rail" output amplifier can give you a false sense of security. One would think that this means the amplifier will operate as an amplifier (and not a comparator) over the full output range. Figure 6.6 illustrates what the output swing of a single-supply amplifier looks like when you drive the output to the rails. In Figure 6.6, it is noticeable that the linearity of the amplifier starts to degrade long before reaching the output swing maximums. If you operate the output of an amplifier beyond the linear region of this curve, the input to output relationship of the signal will be nonlinear.

The conditions of the DC open-loop gain (A_{OL}) specification really define the linear operating output range of the amplifier. The definition of DC open-loop gain (A_{OL}) is:

A_{OL} (dB) = 20 log ($\Delta V_{OUT} / \Delta V_{OS}$)

where V_{OUT} is the output voltage,

V_{OS} is the input offset voltage,

ΔV_{OUT} is ($V_H - V_L$),

When you drive the output high, V_H is the voltage level at the output,

V_{OH} is the maximum voltage level with respect to V_{DD} that the output can reach.

V_L is the voltage level of the output when it is driven low, and

V_{OL} is the maximum voltage level with respect to V_{SS} that the output can reach.

ΔV_{OS} is the change in offset voltage given the change in output voltage,

Also, $V_H < V_{OH}$ and $V_L > V_{OL}$.

Using an A/D converter to capture the output performance of the amplifier (see Figure 6.7), Figure 6.8 illustrates an example of an amplifier's nonlinearity characteristics. Figure 6.9 shows data correcting the amplifier errors.

Figure 6.7 shows a circuit where a CMOS amplifier drives the input of a 12-bit A/D converter. The input common-mode range of the amplifier is zero volts to V_{DD} − 1.2 V. These specifications indicate that the amplifier does not have rail-to-rail input capability. To work around this limitation, the gain of this circuit is +2 V/V. The amplifier circuit uses a gain value of +2 V/V so to not violate the common-mode range limitations of the input stage of the amplifier.

V_{OL} and V_{OH} specify the rail-to-rail output swing characteristics of the amplifier when operating as a comparator. The V_{OL} minimum specification of this amplifier is 15 mV above ground. Given that the amplifier is in a gain of +2 V/V, input signals above 7.5 mV will cause the output will stay 15 mV above ground. The V_{OH} maximum specification of this amplifier is V_{DD} − 20 mV. Assuming a power supply voltage of 5 V = V_{DD}, input signals above 2.29 V will cause the output amplifier to stay 20 mV below the power supply.

The open-loop gain specification conditions are 300 mV < V_{OUT} < (V_{DD} − 300 mV), R_L = 25 kΩ to V_{DD}/2. Given these conditions, amplifier output distortion may occur with input signals outside of 150 mV ($V_{OUT} = V_{SS}$ + 300 mV) to 2.35 V ($V_{OUT} = V_{DD}$ − 300 mV).

Putting the Amp Into a Linear System

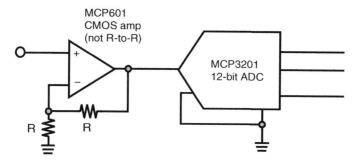

Figure 6.7: This common application uses an amplifier in a gain of +2 V/V to drive an A/D converter. R = 25 kΩ. The common-mode range of the amplifier is from ground to V_{DD} − 1.2 V. The output swing capability of the amplifier ranges from one rail to the other. The V_{OL} minimum specification of this amplifier is 15 mV above ground. The V_{OH} maximum specification of this amplifier is V_{DD} − 20 mV. The conditions of the open-loop gain test is 300 mV < V_{OUT} < (V_{DD} − 300 mV), RL = 25 kΩ to V_{DD}/2. The input range of the A/D converter is zero volts to V_{DD}. The data in Figure 6.8 and Figure 6.9 uses this circuit configuration.

Figure 6.8 shows FFT plots from the circuit in Figure 6.7. (Refer to Appendix B for FFT terminology.) These diagrams show the FFT response of the amplifier/analog-to-digital converter combination to a 1 kHz signal in a 5 V system. It is important to note that the closed-loop bandwidth of the amplifier is approximately 3 MHz (typ) and a slew rate of 2.3 V/μs (typ). The amplifier does not contaminate the signal because of its bandwidth limitations. Rather, the output swing range of the amplifier distorts the signal. In Figure 6.8, the output voltage swing is 140 mV to 4.66 V. In this 5 V supply system, the headroom between the signals and rails is 140 mV.

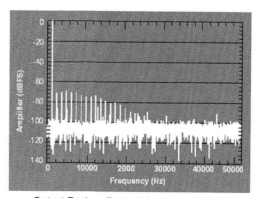

Figure 6.8: This FFT graph shows the digital output of the 12-bit SAR converter, as shown in Figure 6.7. The sampling speed of the converter is 100 ksps. The data represents the FFT calculation of 1024 repetitive samples.

Output Peak-to-Peak of Op Amp = 4.72 V

You can see the distortion in this plot (Figure 6.8). The fundamental frequency is 1 kHz. The distortion occurs at 2 kHz, 3 kHz, 4 kHz and so on. This is pretty nasty. It appears as if the amplifier is distorting the signal.

Chapter 6

Figure 6.9 shows that data where the input signal to the amplifier is less than in Figure 6.8. Because of the reduction in input signal, the output signal is 272 mV from each rail. The data shows considerable improvement.

In Figure 6.9, the amplifier gain and sampling frequency are no different from the conditions of the data in Figure 6.8. The difference between these two figures is the output peak-to-peak signal of the op amp. The distortion that you see in Figure 6.9 is from the A/D converter.

The amplifier V_{OH} and V_{OL} specifications are typically 15 mV to 20 mV from the rails, inclusive. As you can see from these two figures, the output stage of the amplifier is becoming nonlinear before the point where V_{OL} and V_{OH} have an influence.

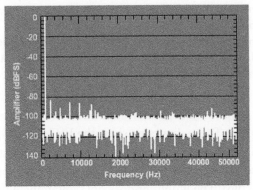

Figure 6.9: The harmonic distortion in this data has improved dramatically, as compared to Figure 6.8. Now the A/D converter, not the amplifier, dominates the distortion.

Output Peak-to-Peak of Op Amp = 4.456 V

The data shown in Figure 6.9 has the MCP3201 (Microchip), 12-bit A/D converter, and MCP602 (CMOS operational amplifier, Microchip) in the circuit. The MCP3201 performance specifications are THD = –82 dB and SNR = 72 dB.

Using the Input and Output Correctly

The typical input bias current of a single-supply bipolar amplifier ranges from a few nano-amperes to hundreds of nano-amperes over temperature. The typical input bias current of CMOS amplifiers range from a few pico-amperes to a hundreds of pico-amperes over temperature (85°C) and thousands of pico-amperes at 125°C. The impact of the error introduced by the input bias current depends on the magnitude of the source resistance and the gain of the circuit. Figure 6.10 shows an example of a circuit that will have an impact due to high input bias current.

The circuit design in Figure 6.10 converts the light energy that impinges on the photodiode (D_P) into charge (or current over time). In the photodiode pre-amp stage, the current from the photodiode flows through the feedback resistor, R_F, generating a voltage at the output of A_1. The output of A_1 is directly connected to R_5, which is a part of a 2nd order low-pass gain/filter

Putting the Amp Into a Linear System

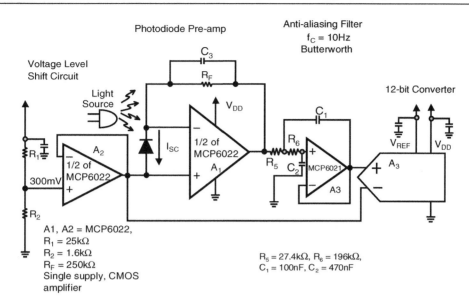

Figure 6.10: In photo detection circuits, the amplifier input bias current could generate voltage drop across the parallel combination of R_F and the photodiode parasitic resistance. In anti-aliasing filter circuits, input bias current can generate a voltage error across input resistors (R_5 and R_6).

stage (anti-aliasing filter). This stage uses A_2, R_5, R_6, C_1 and C_2. The filter sends the output signal to a 12-bit A/D converter.

If the amplifiers in this circuit design are bipolar, the high input bias current from the amplifiers can cause voltage errors in both stages. In the photo detection circuit, the amplifier input bias current (A_1) generates a voltage drop across the R_F. As an example, assume that the feedback resistance (R_F) is equal to 250 kΩ and the input bias current of A_1 is equal to 100 nA. The resulting voltage error from this combination is equal to 25 mV, which appears at the output of the amplifier.

In the anti-aliasing filter circuit (A_3), the input bias current of the amplifier generates a voltage error across the input resistors (R_5 and R_6) at DC. Using the resistor values for R_5 and R_6 equal to 12.9 kΩ and 595 kΩ, respectively, a 100 nA input bias current from the amplifier generates a 61 mV error. This added to 25 mV from the previous stage equals 84 mV of error at the input of the 12-bit A/D converter. If the LSB size of the 12-bit A/D converter is equal to 1.22 mV, this error will produce 69 counts of offset error. You can calibrate out this error, but if dynamic range is an issue, you will lose these codes at the top or bottom.

You can recalculate this example using CMOS amplifiers with an input bias current of 60 pA (over temperature) instead of the bipolar amplifier selected above. With this new amplifier, the voltage error in the first stage is 15 μV. The voltage error generated as a result of the

Chapter 6

two resistors in the anti-aliasing filter stage is 36.5 µV, which is presented to the 12-bit A/D converter input. Now with an A/D converter LSB size of 1.22 mV, the error from the analog front-end produces a 0.0042 bit error. Actually, you don't even see this error at the output of the converter.

The input and output stage requirements of A_1 and A_3 are different. The input range of A_1 is near ground. Biasing the noninverting input to ground does this. For this reason, an amplifier that does not have rail-to-rail input operation is sufficient. The input range of A_3 is much different. The A_1 gains the signal so that the range of the signal into A_3 is nearly rail-to-rail. Consequently, A_3's input stage must have rail-to-rail capability. Since the DC gain of A_3 is unity, the input-stage, crossover-distortion error is relatively small in this 12-bit system.

A Word About the DC Specs: V_{OS}, A_{OL}, CMRR and PSRR

Four DC specifications use the offset voltage of the amplifier in their calculations. These specifications are: Input offset voltage (V_{OS}), open-loop gain (A_{OL}), common-mode rejection ratio (CMRR), and power supply rejection ratio (PSRR). Each one of these specifications is important. But, your job is to try to keep them in perspective as you go down the specmanship lane. Specmanship lane is the place where we make decisions about devices from the data sheet alone with little regard to the actual application.

Input offset voltage defines the maximum voltage difference that will occur between the two input terminals in a closed-loop circuit, while the amplifier is operating in its linear region. The units of the room temperature specification are µV or mV. Over temperature, the units are µV/°C, as well as absolute values of µV or mV. Offset voltage is a voltage source at the noninverting input of the amplifier, as shown in Figure 6.11

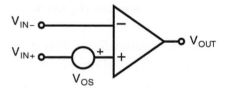

Figure 6.11: The best place to model the offset voltage of an amplifier is at the front end of the noninverting input.

The offset voltage can vary from part to part and with temperature, as shown in the distribution graphs in Figure 6.12. The offset voltage of a particular amplifier does not vary unless the temperature, power-supply voltage, common-mode voltage or output voltage changes. The graphs in Figure 6.12 illustrate this. The offset voltage error of a particular amplifier may or may not be a problem, depending on the application circuit. For example, an amplifier in a buffer configuration that has larger offset voltage errors (2 mV to 10 mV) doesn't create significantly different performance than high precision amplifiers with extremely low offset voltage specifications, (100 µV to 500 µV). On the other hand, amplifiers with a high offset-voltage that are in a high, closed-loop gain configuration can dramatically compromise the dynamic range of the circuit.

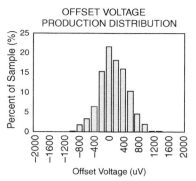
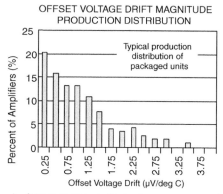

a. Room temperature offset voltage measurements of several MCP601 (Microchip Technology)

b. Over temperature offset voltage measurements of several MCP601 (Microchip Technology)

Figure 6.12: The offset voltage of several devices from a production run (a) can vary from device to device. The same is true for the offset voltage change over temperature (b).

For example, the design in the Figure 6.13 produces an output voltage of:

$$V_{OUT} = (1 + R_F / R_{IN})(V_{IN} + V_{OS})$$

Unfortunately, the gain factor of the amplifier circuit multiplies offset voltage as well as the input signal with the gain constant. In this example, $(1 + R_F / R_{IN})$ is equal to 101 V/V. An amplifier with an offset voltage of 1 mV would produce a constant DC error at the output of 101 mV. In a 5 V system, 101 mV lessens the dynamic range by approximately 2%.

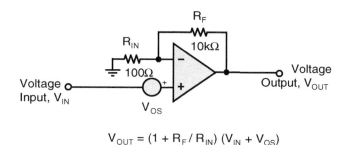

$$V_{OUT} = (1 + R_F / R_{IN})(V_{IN} + V_{OS})$$

Figure 6.13: The gain of the offset voltage and signal are the same, $1 + R_F/R_{IN}$.

I have always said to the customers that I talk to and the designers that I work with, that A_{OL}, CMRR, and PSRR should be respectable. This means that you shouldn't focus on these parameters. You should just make sure that they are within reason (≥ 80 dB). These three specifications won't break the bank, but the offset voltage error may.

A working definition of A_{OL} is:

$$A_{OL} = 20 \log (\Delta V_{OUT} / \Delta V_{OS})$$

Chapter 6

You may object to this simplification, but I chose to make this specification easy to understand and test. This formula implies that the common-mode voltage of the input stage is static, and the power supply voltages are constant. It also takes the difference between two offset voltage measurements. You will find that if an amplifier has high offset voltage, it does not imply the open-loop gain will be lower.

You can use the circuit in Figure 6.14 to test this parameter.

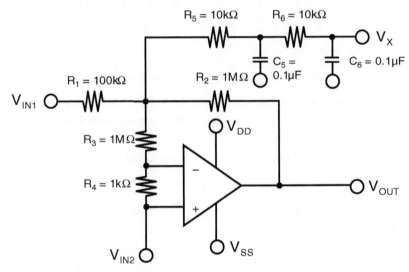

Figure 6.14: This circuit measures the DC parameters of V_{OS}, A_{OL}, CMRR and PSRR.

The offset voltage in this circuit is equal to:

$$V_{OS} = (V_X - V_{IN2})/(1 + R_3/R_4)$$

V_{IN1} varies the output of the amplifier; V_{IN2} is kept constant and equal to $(V_{DD} - V_{SS})/2 + V_{SS}$; V_{SS} and V_{DD} are kept constant in relation to V_{IN2}, which is the common-mode voltage of the amplifier.

The impact of your open-loop gain error affects your circuit when the amplifier output changes. If the output of your amplifier changes (because of an input voltage change), the offset voltage will adjust to accommodate the open-loop gain error. So how much offset voltage error can you tolerate? If you use your amplifier in a buffer configuration, I would guess that you could tolerate a great deal of offset error. In fact, the amplifier's V_{OS} will out-shadow the open-loop gain error.

Putting this in real terms, if the open-loop gain of your amplifier is 80 dB (which is fairly low), and the output swing of the amplifier is from 500 mV to 4.5 V, the change in offset voltage will be 0.4 mV. The offset voltage of your amplifier could easily be several millivolts.

This amplifier characteristic starts to have an impact on your circuit as you increase the closed-loop gain.

Let's take a stab at the CMRR specification. A working definition of CMRR is:

$$\text{CMRR} = 20 \log (\Delta V_{CM} / \Delta V_{OS})$$

Does this look familiar? You can bench test this parameter with the circuit in Figure 6.14. As you test it, ensure that the output and power supply voltages remain constant. The only variable in this test is the common-mode voltage of the amplifier (V_{IN2}).

Keeping the output of the amplifier constant, in relation to V_{DD} and V_{SS}, by changing V_{IN1} properly runs this test. V_{IN2} changes the common-mode voltage. V_{IN2} ranges from V_{SS} to V_{DD} if the device has a composite input stage. Otherwise, the high side or low side is limited. V_{SS} and V_{DD} are constant in relation to V_{OUT}.

Let's again put this specification into perspective. If the CMRR performance of your amplifier is 80 dB (or not very good) and you use the full input range of the amplifier, the offset voltage change as the common-mode voltage spans for V_{DD} to V_{SS} is 0.4 mV.

Finally, a working definition of the PSRR specification is:

$$\text{PSRR} = 20 \log (\Delta V_{POWER\ SUPPLY} / \Delta V_{OS})$$

You can bench test this parameter with the circuit in Figure 6.14. This time, as you test, ensure that the output and common-mode voltages are constant and the power supply voltages are simultaneously changed. If the output and common-mode voltage start at half-way between the power supply voltages, this relationship has to continue.

If you go through the math, you will find that the PSRR has very little effect on amplifier errors. In fact, it is very unusual for an application circuit to experience any more than a few hundred millvolt change in the power supply.

Every Amplifier is Waiting to Oscillate, and Every Oscillator is Waiting to Amplify

What is operational amplifier circuit stability and how do you know when you are on the hairy edge or near oscillation? Typically, there is a feedback system around the op amp to stabilize the gain variability from part to part. With this approach, the stability of your amplifier circuit depends on the accuracy of the resistors in your circuit, not your op amp. Using resistors around your op amp provides circuit "stability." At least you hope that a predictable gain is ensured. However, it is possible to design an amplifier circuit that does quite the opposite.

You can design an amplifier circuit that is extremely unstable to the point of oscillation. In these circuits, the closed-loop gain is somewhat trivial because an oscillation is swamping out your results at the output of the amplifier. In a closed-loop amplifier system, stability can

Chapter 6

be ensured if you control the phase margin of the amplifier system. In this evaluation, the Bode stability analysis technique is commonly used. With this technique, the magnitude (in decibels, dB) and phase response (in degrees, °) of both the open-loop response of the amplifier and circuit feedback factor are included in the Bode plot. This section of the chapter will look at these concepts and make suggestions on how to avoid the design of a "singing" circuit when you actually want frequency stability as a primary objective.

The Internal Basics of the Operational Amplifier Block Diagram

Before we get going on the frequency analysis of an amplifier circuit, we need to review a few amplifier topology concepts. Figure 6.15 shows the critical internal op amp elements that you will need to be familiar with if you engage in a frequency analysis. This amplifier has five terminals, as expected, but it also has parasitics, such as input capacitance and a frequency dependent open-loop gain.

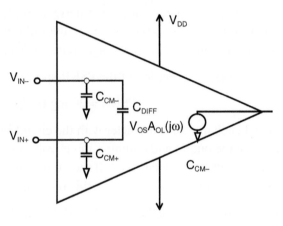

Figure 6.15: The voltage feedback operational amplifier frequency model includes the input capacitances (C_{DIFF} and C_{CM}) to ensure that the interaction of the external input source parasitics and the feedback parasitics can be taken into account in a frequency evaluation. This model also has the internal open-loop gain over frequency ($A_{OL}(j\omega)$). These two parameters ensure that the internal parasitics of the output stage are in the analysis.

The two input terminals have common-mode capacitances to ground (C_{CM}) and differential capacitance (C_{DIFF}) between the inputs. $A_{OL}(j\omega)$ represents the open-loop gain, frequency response of the amplifier. Figure 6.16 shows the frequency response of a typical voltage feedback amplifier using the Bode plot method.

The amplifier gain response over frequency ($A_{OL}(j\omega)$) for a voltage feedback amplifier is usually modeled with a simple 2nd order transfer function. This 2nd order transfer function has two poles. The two plots in Figure 6.16 illustrate the gain (top) and phase response (bottom) of a typical operational amplifier. The units of the y-axis of the gain curve are decibels (dB).

Ideally, the open-loop gain of an amplifier is equal to the magnitude of the ratio of the voltage at the output terminal divided by the difference of the voltages applied between the two input terminals.

$$A_{OL} \text{ (dB)} = 20\log |(V_{OUT} / (V_{IN+} - V_{IN-}))|$$

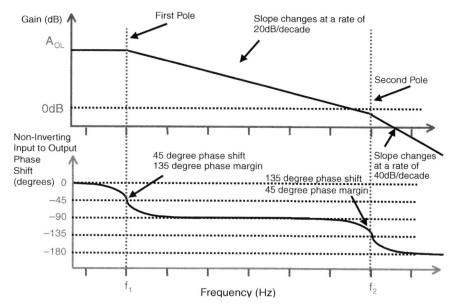

Figure 6.16: A gain plot and phase plot illustrates the frequency behavior of a voltage-feedback amplifier. In this simple representation, the amplifier has two dominant poles. The first pole occurs at lower frequencies, typically between 10 Hz to 1 kHz (depending on the gain-bandwidth product of the amplifier). The second pole resides at higher frequencies. This pole occurs at a higher frequency than the zero dB frequency. If it is lower than this frequency, the amplifier is usually unstable in a unity gain circuit.

It would be nice if this open-loop gain ratio were infinite. But in reality, the complete frequency response of the open-loop gain, $A_{OL}(j\omega)$, is less than ideal at DC. The open-loop, gain curve attenuates at a rate of 20 dB/decade. This attenuation starts at the frequency where the first pole in the transfer function appears. This is illustrated in the Bode plot in Figure 6.16.

Usually, the first pole of the open-loop response of an operational amplifier occurs between 1 Hz to 1 kHz. The second pole occurs at a higher frequency, shortly after the frequency where the open-loop gain curve crosses 0 dB. The gain response of an amplifier starts to fall off at 40 dB/decade at the frequency where the second pole occurs.

The units of the y-axis of the phase plot in Figure 6.16 (bottom plot) are degrees. You can convert degrees to radians with the formula:

$$\text{Phase in radians} = (\text{Phase in degrees}) / 360°$$

Phase in degrees can be translated to phase delay or group delay (seconds), with the formula:

$$\text{Phase delay} = -(\delta\text{phase} / \delta f)/360°$$

The same x-axis frequency scale aligns across both plots.

Chapter 6

The phase response of an amplifier in this open-loop configuration is also predictable. A phase shift is the change in the phase of the noninverting input versus output as compared to DC. A phase margin is the difference between the actual phase of the noninverting input versus output and –180°. The phase-shift or change from the noninverting input to the output of the amplifier is 0° at DC. Therefore, the phase-shift from the inverting input terminal to the output is equal to –180° at DC.

At one decade ($1/10\ f_1$) before the first pole, f_1, the phase relationship of noninverting input to output has already started to change (~ –5.7°). At the frequency where the first pole appears in the open-loop gain curve (f_1), the phase margin has dropped to –45°. The phase continues to drop for another decade ($10f_1$) where it is 5.7° above its final value of –90°. These phase response changes are repeated for the second pole, f_2.

What is important to understand are the ramifications of changes in this phase relationship with respect to the input and output of the amplifier. One frequency decade past the second pole, the phase-shift of the noninverting input to output is ~ –180°. At this same frequency, the phase-shift of the inverting input to output is zero or ~ –360°. With this type of shift, V_{IN+} is actually inverting the signal to the output. In other words, the roles of the two inputs have reversed. If the role of either input changes like this, the amplifier will ring as the signal goes from the input to the output in a closed-loop system. The only thing stopping this condition from occurring with the stand-alone amplifier is that the gain drops below 0 dB. If the open-loop gain of the amplifier drops below 0 dB in a closed-loop system, the feedback is essentially "turned-off."

Stability in Closed-Loop Amplifier Systems

Typically, operational amplifiers have a feedback network around them. This reduces the variability of the open-loop gain response from part to part. Figure 6.17 shows a block diagram of this type of network.

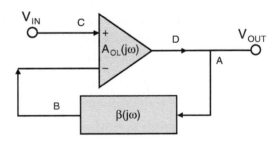

Figure 6.17: A block diagram of an amplifier circuit which includes the amplifier gain cell, A_{OL}, and the feedback network, b.

In Figure 6.17, $\beta(j\omega)$ represents the feedback factor. Due to the fact that the open-loop gain of the amplifier (A_{OL}) is relatively large and the feedback factor is relatively small, a fraction of the output voltage is fed back to the inverted input of the amplifier. This configuration sends the output back to the inverting terminal creating a negative-feedback condition. If β were fed back to the noninverting terminal, this small fraction of the output voltage would be added instead of subtracted. In this configuration, there would be positive feedback and the output would eventually saturate.

Closed-Loop Transfer Function

If you analyze the loop in Figure 6.17, you must assume an output voltage exists. This makes the voltage at "A" equal to $V_{OUT}(j\omega)$. The signal passes through the feedback system, $\beta(j\omega)$, so that the voltage at "B" is equal to $\beta(j\omega)V_{OUT}(j\omega)$. The voltage, or input voltage, at "C" is added to the voltage at "B". "C" is equal to $(V_{IN}(j\omega) - b(j\omega)V_{OUT}(j\omega))$. With the signal passing through the gain cell, $A_{OL}(j\omega)$, the voltage at point "D" is equal to $A_{OL}(j\omega)(V_{IN}(j\omega) -\beta(j\omega)V_{OUT}(j\omega))$. This voltage is equal to the original node, "A", or V_{OUT}. The formula that describes this complete closed-loop system is equal to:

$$"A" = "D" \text{ or}$$
$$V_{OUT}(j\omega) = A_{OL}(j\omega)(V_{IN}(j\omega) -\beta(j\omega)V_{OUT}(j\omega))$$

By collecting the terms, the manipulated transfer function becomes:

$$V_{OUT}(j\omega)/V_{IN}(j\omega) = A_{OL}(j\omega) /(1 + A_{OL}(j\omega)\, \beta(j\omega))$$

This formula is essentially equal to the closed-loop gain of the system, or $A_{CL}(j\omega)$.

This is a very important result. If the open-loop gain ($A_{OL}(j\omega)$) of amplifier is allowed to approach infinity, the response of the feedback factor can easily be evaluated as:

$$A_{CL}(j\omega) = 1/\beta(j\omega)$$

This formula allows an easy determination of the frequency stability of an amplifier's closed-loop system.

Calculation of 1/β in an Amplifier Circuit

The easiest technique to use when calculating 1/β is to place a source directly on the noninverting input of the amplifier and ignore error contributions from the amplifier. One could argue that this calculation will not give the appropriate circuit, closed-loop gain, equation for the actual signal and this is true. However, if you use this calculation, you can determine the level of circuit stability.

You can calculate 1/β by using the circuits in Figure 6.18.

In Figure 6.18a and 6.18b, $V_{STABILITY}$ is a fictitious voltage source, equal to zero volts. This voltage is used for the 1/β stability-analysis. Note that this source is not the actual application input source.

Chapter 6

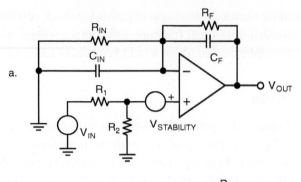

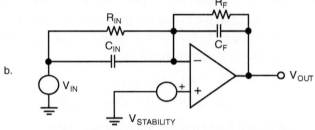

Figure 6.18: The input signal in circuit (a) at DC will be gained by $(R_2/(R_1 + R_2))(1 + R_F/R_{IN})$. The input signal in circuit (b) has a DC gain of $-R_F/R_{IN}$. Neither of these gain equations match the DC gain of the feedback factor, $1/\beta$.

If the open-loop gain of the amplifier is infinite, the transfer function of this circuit is equal to:

$$V_{OUT}/V_{STABILITY} = 1/\beta$$

$$1/\beta = 1 + (R_F\|C_F)/(R_{IN}\|C_1)$$

($C_1 = C_{IN} + C_{CM-}$ for Figure 6.18a, and $C_1 = C_{IN} + C_{CM-} + C_{DIFF}$ for Figure 6.18b)

or,

$$1/\beta(j\omega) = (R_{IN}((j\omega)R_F C_F + 1) + R_F((j\omega)R_{IN} C_1 + 1))/ R_{IN}((j\omega)R_F C_F + 1)$$

In the above equation, when ω is equal to zero:

$$1/\beta(j\omega) = 1 + R_F/R_{IN}$$

As ω approaches infinity,

$$1/\beta(j\omega) = 1 + C_1/C_F$$

The transfer function has one zero and one pole. The zero is located at:

$$f_Z = 1/(2\pi R_{IN}\|R_F(C_1 + C_F))$$

$$f_P = 1/(2\pi R_F C_F)$$

The Bode plot of the $1/\beta(j\omega)$ transfer function of the circuit in Figure 6.18a is shown in Figure 6.19.

Putting the Amp Into a Linear System

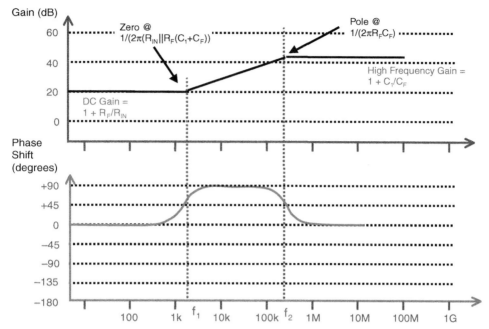

Figure 6.19: These are the Bode plots of the inverse of the feedback factor (1/β) for the circuit in Figure 6.18a using $V_{STABILITY}$ as the input source.

Once again, in Figure 6.18b, you use the input source for this analysis. This input source is not the same as the input source for the actual application circuit. However, the amplifier stability is determined in the same manner. The closed-loop transfer function, using $V_{STABILITY}$ is equal to:

$$V_{OUT}/V_{STABILITY} = 1/\beta$$

$$1/\beta = 1 + (R_F \| C_F)/(R_{IN} \| C_{IN})$$

Note that the transfer functions of 1/β between Figure 6.18a and Figure 6.18b are identical.

Determining System Stability

If you know the phase margin, you can determine the stability of a closed-loop amplifier system. In this analysis, the Bode stability analysis technique is commonly used. With this technique, the magnitude (in dB) and phase response of both the open-loop response of the amplifier and circuit feedback factor are included in a Bode plot.

The system closed-loop gain is equal to the lesser (in magnitude) of the two gains. The phase response of the system is the equal to the open-loop gain phase-shift minus the inverted feedback factor's phase-shift. The closed-loop bandwidth is equal to the frequency where the open-loop gain curve crosses the 1/β curve.

Chapter 6

You define the stability of the system at the frequency where the open-loop gain of the amplifier intercepts the closed-loop gain response. At this point, the theoretical phase-shift of the system should be greater than –180°. In practice, the system phase-shift should be smaller than –135°. Figures 6.20 through 6.23 illustrates this technique. The cases presented in Figures 6.20 and 6.21 represent stable systems. The cases presented in Figures 6.22 and 6.23 represent unstable systems.

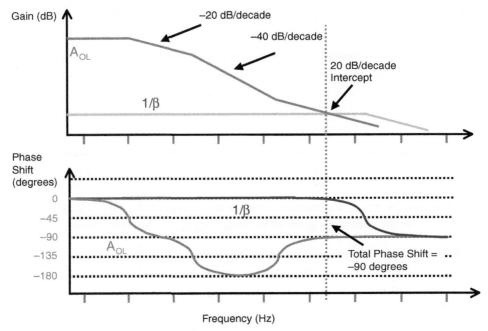

Figure 6.20: This closed-loop system is stable with a phase-shift of –90° at the intercept of the A_{OL} and 1/β curves.

In Figure 6.20, the open-loop gain of the amplifier ($A_{OL}(j\omega)$) starts with a zero dB change in frequency and quickly changes to a –20 dB/decade slope. At the frequency where the first pole occurs, the phase-shift is –45°. At the frequency one decade above the first pole, the phase-shift is approximately –90°. As the gain slope progresses with frequency, a second pole is introduced, causing the open-loop gain response to change –40 dB/decade. Once again, this is accompanied with a phase change. The third incident that occurs in this response is where a zero is introduced, and the open-loop gain response returns back to a –20 dB/decade slope.

The 1/β curve in this same graph starts with a zero dB change with frequency. 1/β remains flat with increased frequency until the very end of the curve where a pole occurs and the curve starts to attenuate –20 dB/decade.

The point of interest in this graph is where the $A_{OL}(j\omega)$ curve intersects the 1/β curve. The rate of closure of 20 dB/decade between the two curves suggests the phase margin of the system, and in turn, predicts the stability. In this situation, the amplifier is contributing a −90° phase-shift and the feedback factor is contributing a 0° phase-shift. The phase-shift, and consequently, the stability of the system are determined at this intersection point. The system phase-shift is calculated by subtracting the 1/β(jω) phase-shift from the A_{OL}(jω) phase-shift. In this case, the system phase-shift is −90°. Theoretically, a system is stable if the phase-shift is between zero and −180°. In practice, you should design to a phase-shift of −135° or smaller.

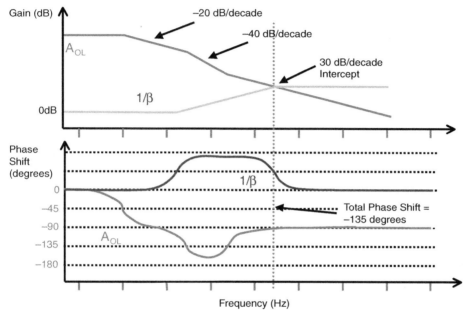

Figure 6.21: This system is marginally stable with a −135° phase-shift at the intersection of the two gain curves.

In the case presented in Figure 6.21, the point of intersection between the A_{OL}(jω) curve and the 1/β(jω) curve suggests a marginally stable system. At that point, the A_{OL}(jω) curve is changing −20 dB/decade. The 1/β(jω) curve is changing from a +20 dB/decade to a 0 dB/decade slope. The phase-shift of the A_{OL}(jω) curve is −90°. The phase-shift of the 1/β(jω) curve is +45°. The system phase-shift is equal to −135°.

Although this system appears to be stable—that is, the phase-shift is between 0° and −180°, circuit implementation will not be as clean as calculations or simulations would imply. Parasitic capacitance and inductance on the board can contribute additional phase errors. Therefore, this system is "marginally stable" with this magnitude of phase-shift. This closed-loop circuit has a significant overshoot and ringing with a step response.

In Figure 6.22, the A_{OL}(jω) is changing at a rate of −20 dB/decade. The 1/β(jω) is changing at a rate of +20 dB/decade. The rate of closure of these two curves is 40 dB/decade and the system phase-shift is −168°. The stability of this system is very questionable.

Chapter 6

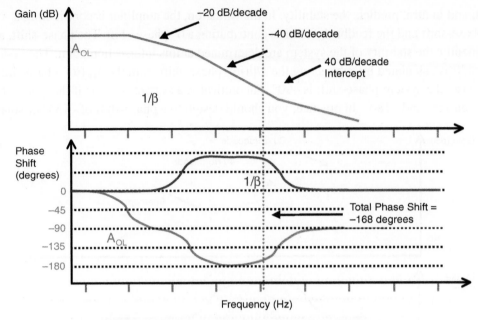

Figure 6.22: In a practical circuit implementation, given layout parasitics, this system is unstable.

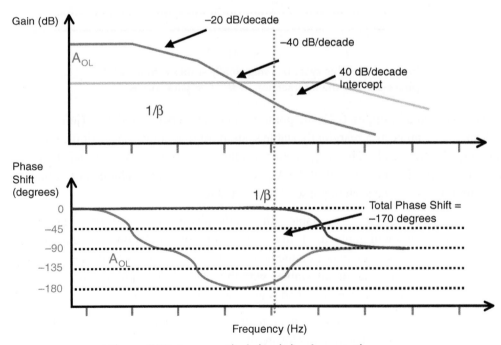

Figure 6.23: In a practical circuit implementation, given layout parasitics, this system is also unstable.

Putting the Amp Into a Linear System

In Figure 6.23, $A_{OL}(j\omega)$ is changing at a rate of -40 dB/decade. The $1/\beta(j\omega)$ is changing at a rate of 0 dB/decade. The rate of closure of these two curves is 40 dB/decade indicating a phase-shift of $-170°$. The stability of this system is also questionable.

Summarizing Stability in the Frequency Range

At the beginning of this section, I asked, "What is operational amplifier circuit stability and how do you know when you are on the hairy edge?" There are many definitions of stability in analog, such as unchanging over temperature, unchanging from lot to lot, noisy signals and so forth. However, an analog circuit becomes critically unstable when to output unintentionally oscillates without excitation. This kind of stability problem will stop your progress with your circuit design until you track it down.

You can only evaluate this kind of stability in the frequency domain. A quick paper and pencil examination of your circuit will readily provide the insight into your oscillation problem. The relationship between the open-loop gain of your amplifier and the feedback system over frequency will quickly identify the source of the problem. If you use gain and phase Bode plots, you will be able to estimate where these problems reside. You will minimize any ringing if you keep the closed-loop phase shift below $-135°$. If you do this work up front with amplifiers, you can avoid those dreadful designs that kick into an unwanted song, a.k.a., "The Amplifier Circuit Blues."

Time Domain Performance

The time domain responses of amplifier circuits provide a real-world result of the previous frequency discussions. Figure 6.24 shows the graphical definition of time domain specifications. The waveform in this figure depicts the response of the output of the amplifier with regards to a step-response at the input of the circuit. I will refer to this figure throughout the following discussion.

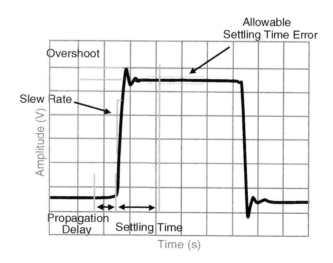

Figure 6.24: The time domain step response behavior of an amplifier configured in a closed-loop system uses the definitions diagrammed here.

Chapter 6

Slew Rate (SR)

The slew rate (SR) specification for an amplifier quantifies the speed at which the output terminal can execute a full-scale input to output voltage swing. The amplifier controls slew rate internally. The internal tail currents of the amplifier charge and discharge internal capacitors. The units of this specification are in volts per second. You measure slew rate from 10% to 90% at the output of the amplifier, through the full-scale voltage swing. The most challenging amplifier circuit for this type of specification is the buffer or follower configuration, as discussed in Chapter 5. In this configuration, you pull the input terminals out of their nonlinear region by virtue of the fact that the inverting input terminal is connected directly to the slow moving output terminal.

Settling Time (t_s) and Overshoot

At the top or bottom of the full-scale slew transition, a degree of ringing occurs. This ringing is directly related to the phase-shift of the closed-loop system and described in terms of overshoot and the amount of time before the signal to settles, within a specified error band. The overshoot occurs at the highest peak in this portion of the waveform, which occurs at the beginning of the ringing of the system. The magnitude of the system overshoot and the amount of time required for the system to settle directly relates to the frequency domain phase-shift of the system.

The settling time (t_s) of an amplifier circuit is defined as the amount of time that is required for the output of the amplifier to slew and then be confined to the defined error band. This time starts at the point where the output first responds to the input excitation until the last occurrence of the output signal being outside the error band, as illustrated in Figure 6.24.

A settling time error is most noticeable in applications where the common-mode input of the amplifier has a full-scale step-response applied. In this situation, the amplifier goes into a full slew condition (SR) and then settles to its final value (t_s). The amplifier in the circuit

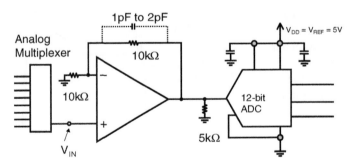

Figure 6.25: When the multiplexer switches from one channel to the next, it is possible that the amplifier receives a stepped signal. Figure 2.26 shows the step response of this amplifier circuit. The amplifier circuit does not make the system unstable. However, you should delay the A/D converter conversion while the op amp circuit is settling to final value.

in Figure 6.25 is a perfect candidate for this type of signal. Each signal at the input of the multiplexer in this system could be fast or slow moving, or even DC. When the multiplexer switches, it presents a step response to the amplifier.

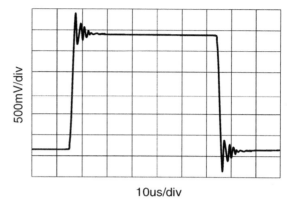

Figure 6.26: The phase-shift of the amplifier in Figure 6.25 shows up in the step response in terms of overshoot (15%) and ringing.

Go Forth

I hope that I have brought up the most common, AC and DC problems with op amp designs in this chapter. You, at least, will have a place to start when you get bizarre behavior out of your amplifier. The best place to look for DC errors is in the input stage and/or the output stage. The input stage topology can prevent you from using an amplifier in a buffer configuration. But most likely, the input stage (if it is rail-to-rail) will generate a degree of distortion to the offset voltage.

This distortion will probably not cause as much heartache as the output stage. The output stage of any amplifier will not go all the way to the rail. Worse yet, nonlinearities (or distortion) will creep as the amplifier goes towards the rail.

The difference between the A_{OL} conditions and an output swing condition are profoundly different. The A_{OL} specification validates the voltage-output swing-test by implying that the operational amplifier is operating within its linear region. However, taking this discussion beyond the difference of these specifications, the output of your single-supply amplifier will never reach the power supply rail. Design your circuits accordingly.

The stability of an amplifier circuit is straightforward as long as you take the real capacitors, the amplifier capacitors, and the layout capacitors into consideration. A Bode plot analysis helps identify problem areas very quickly. You may say that this graphical method is not accurate, and you are right. However as a first pass, the Bode plot analysis is very revealing as long as you know what you are looking for. Always examine the plot at the intersection of the open-loop gain curve and the feedback curve for an estimated rate of closure and phase shift.

Chapter 6 References

Introduction to Operational Amplifier Theory and Applications, Wait, Huelsman, Korn, McGraw Hill, 1975.

Noise Reduction Techniques in Electronic Systems, Ott, Henry W., John Wiley & Sons, 1988.

Intuitive Operational Amplifiers, Frederiksen, Thomas M., McGraw Hill, 1988.

Design with Operational Amplifiers and Analog Integrated Circuits, Sergio Franco, McGraw Hill.

Analog Circuit Design, Williams, Jim, Butterworth-Heinemann.

"Tuning In Amplifiers," Baker, Bonnie C., AB-105, Burr-Brown Corporation (Texas Instruments).

"Operational Amplifiers: 6 Part," Baker, Bonnie C., First Published in *analogZone* (2002, 2003) and reproduced with permission.

CHAPTER 7

SPICE of Life

CHAPTER 7

SPICE of Life

They say a computer-based simulation of your analog circuit is important. This is because the use of your preferred computer Simulation Program with Integrated Circuit Emphasis (SPICE) program can reduce initial errors and development time. If you use your SPICE simulator correctly, you can drum out circuit errors and nuances before you go to your breadboard. In this manner, you will verify your design before you spend the time to solder your circuit. SPICE helps troubleshoot bench problems; it is a great place to try out different hypotheses. It is also great at "what if" scenarios (for example, exploratory design).

You can view the results from these software tools on a PC with user-friendly, graphical user interface (GUI) suites. This tool will fundamentally provide DC operating (quiescent) points, small signal (AC) gain, time domain behavior and DC sweeps. At a more sophisticated level, it will help you analyze harmonic distortion, noise power, gain sensitivity and perform pole-zero searches. This list is not complete, but generally, SPICE software manufacturers have many of these fundamental features available for the user. By finessing the Monte Carlo and worst-case analysis tools in SPICE, you can predict the yields of your final product. If you use your breadboard for this type of investigation, it could be very expensive and time consuming. All of these things will speed up your application circuit time-to-market.

But, beware. You can effectively evaluate analog products if your SPICE models or macromodels are accurate enough for your application. The key words here are "accurate enough." Such models, or macromodels, should reflect the actual performance of the component without carrying the burden of too many circuit details. Too many details can lead to convergence problems and extremely long simulation times. Not enough details can hide some of the intricacies of your circuit's performance. Worse yet, your simulation, whether you use complete models or just macromodels, may give you a misrepresentation of what your circuit will really do. Remember that a SPICE simulation is simply a pile of mathematical equations that, hopefully, represent what your circuit will do. In essence, a computer product produces imaginary results.

So you might ask, "Why bother?" Is a SPICE simulation worth the time and effort? A pop quiz will help you clarify this question. The circuit in Figure 7.1 shows a fundamental, basic circuit. Is this circuit stable or does it oscillate? Would the output of the amplifier have an unacceptable ring? I would think that you would quickly look at this and say, "That is a silly question. Of course it is stable!" But then again, if you are always looking for the trick question you may be suspicious. So what is the answer?

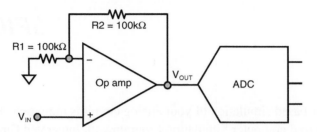

Figure 7.1: A variety of applications throughout the industry have this simple sub-circuit embedded in the system. This circuit simply takes an analog input signal and gains that signal to the output of the amplifier. For instance, an input signal of +1 V_{DC} would become a +2 V_{DC} signal at V_{OUT}. The question is, would this DC signal oscillate? Or, would a 50 kHz sinusoidal signal oscillate or ring? The bandwidth of this amplifier is 2.8 MHz.

This simple amplifier circuit uses an amplifier in a gain of +2 V/V. The amplifier has a 100 kΩ resistor connected to its inverting-input to ground, and 100 kΩ resistor in the feedback loop. It would be easy to assume that this circuit is stable. However, tedious calculations will verify that this amplifier circuit will ring. This is due to the parasitic capacitances around the resistors and the high differential/common-mode capacitance of the amplifier's input stage. For this particular amplifier, the input common-mode capacitance is 6 pF and the differential-mode capacitance is 3 pF. These capacitances interact with the feedback resistor causing a semi-unstable condition. If you bench-test this circuit, you will immediately see this condition on the oscilloscope. Parasitics on the breadboard will aggravate this instability.

Figure 7.2: By enhancing the circuit diagram in Figure 7.1 with the parasitic capacitances of the resistors and amplifier, a simple of a circuit is not so simple. In the DC domain these capacitors will operate as open circuits. In the AC domain, the capacitors will affect the perfect square wave from the input to output. The perfect square wave will have quite a ring at the V_{OUT} node.

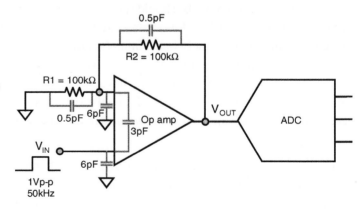

For example, the 100 kΩ resistor in the feedback loop of the amplifier will also have approximately a parallel 0.5 pF parasitic capacitor (see Figure 7.2). This parasitic capacitance on the breadboard to ground could be as high as 2 pF or 3 pF. If you use the amplifier's SPICE macromodel, with input impedances in the model and board parasitics, you will see this problem immediately in your simulation. If you breadboard the circuit, you most certainly will see this ringing.

Changing the values of the two resistors in this circuit solves this problem. Hand calculations will help you find the correct values. A SPICE simulation will facilitate the process. This is a little easier than swapping out resistors on the breadboard until you find the right values. In SPICE, you can also look at the response of the amplifier using various resistors. This will help you find the "corner" of this oscillation. If you go back and change both values to 10 kΩ, you will have great success in SPICE and on the bench.

Figure 7.3. shows the simulation result.

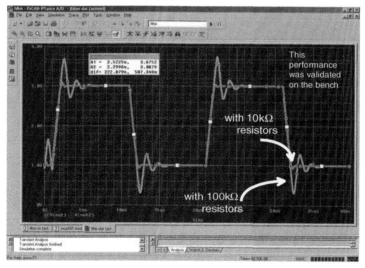

Figure 7.3: You can quickly verify that this simple circuit will ring using a SPICE simulation. If you need to double-check this with a breadboard circuit, that is also a good idea, however, reducing the 100 kΩ resistors down to 10 kΩ resistors solves the problem. You do need to understand where the problem came from before you continue with your circuit design. But this simulation caught a significant stability problem. This ringing problem was an easy one to miss by inspection of the schematic.

In this chapter, we will discuss how to best determine if your SPICE simulation is telling you the truth. We do this by using three techniques. First, we will go through a short list of a few rules of thumb, which will help you examine the validity of your simulations; second, we

Chapter 7

will use common sense (at least where your circuit is concerned) when you first examine the results of your simulation; third, we will engage in an overview of what a macromodel can (or can't) do for you. With this arsenal, you will be able to effectively use a SPICE simulation to weed out most of your circuit problems.

I'll center the discussion on signal quality operational amplifiers. This is only because an amplifier is embedded somewhere in most purely analog circuits. I am going to leave it up to you to explore the remainder of analog macromodels, such as instrumentation amplifiers, difference amplifiers, references and so forth.

What won't we cover in this chapter? I don't intend to give you tips on how to use your favorite SPICE simulator tool. I'll leave that topic up to your SPICE vendor and the numerous books on this topic. I also will not attempt to think for you by giving you cookbook answers to your problems. Rather, I am going to ask you to think through things yourself. Basically, you need to ask, "What do you expect from your circuit? Does your SPICE simulation match your expectations? Why or why not?"

The naysayers in the industry will tell you that your computer-based simulation tools will not work and using them will be a waste of time. These people are a bit misguided, and in my opinion have a superficial view of what this tool can really do. Sure, SPICE tools can lead you astray. But, like any tool, it is only as good as the user. Any insight that you gain from your simulations are provided if your SPICE tools are understood and used properly. Better yet, SPICE simulations will point out problems that you had never anticipated. In most cases, they use double-precision calculations. This makes it easier to detect low-level problems that are impossible to find on the bench. But step back and look at what you have. You very likely will have a simulation circuit that is built out of macromodels that are generated by various manufacturers of the products you are interested in using in your circuit.

The questions that bear asking are, "Does my model or macromodel simulate over temperature. Distortion? AC spec? Over process? Am I expecting the model to simulate these parameters? And to what degree of accuracy? What information do the macromodels I am using really provide?" The only way you can answer these questions is to have a feel for what your circuit will do (in real life), and ask challenging questions about your SPICE simulation results.

To give you a little taste of my experiences with SPICE, during one of my many lives, I was an analog IC designer. As a designer, I was using a transistor in an unorthodox manner. Mind you, my transistor models were the best that we had. My models were full-blown transistor-level models, designed to simulation the accurate behavior of my amplifier design. I was not using a macromodel substitution, but the "real" thing. I believed that a certain configuration of a transistor in the output stage would allow me to design a single-supply amplifier that could go nearly (within a few millivolts) to the negative rail. Although this type of circuit operation is not new today, in 1990 it had a degree of innovation. I arrived at this belief by examining and thinking through the circuit operation on paper. It was a very cool idea!

The simulation told me that there was no way my amplifier would even go near the negative rail. Therefore, I questioned my paper calculations and then the SPICE models. This exercise did not produce any answers, so I finally built the circuit on a breadboard. The tests from the breadboard circuit proved that my paper calculations were correct. With this verification, I went back and created a special model for the culprit transistor in my circuit. I went to the bench to create the new transistor model, using a single transistor in a TO-99 package. This model was required because I was using the transistor in an unorthodox configuration. With those transistor changes, my amplifier circuit model did simulate the circuit accurately. At the "end-of-the-day," I had all three tools in agreement. Modifying the transistor model did this. The problem could have been in my calculations or on the bench, but with these three votes, I was certain that I had a winner. The required three votes were the hand calculations, SPICE simulation and breadboard.

You probably will not dig into your circuits to this level of detail, however, the three steps to a successful, expeditious circuit design still hold true. These steps are: 1) Draw up the concept on paper, 2) Simulate your circuit to your satisfaction, 3) Breadboard critical portions of the circuit. All three of these steps are critical. Your SPICE simulations will not replace any of these steps; it will improve the likelihood of your success.

All of these steps were necessary:

1. Imagine what the circuit will do;
2. Simulate the circuit to match what you imagine it should do;
3. Breadboard the circuit and ensure the pencil design and simulation match. You need all three for a good level of confidence;
4. Adjust the model (or circuit) to match the three conditions (imagined operation, simulated operation and actual operation);
5. Always build a prototype.

You might ask me why I bothered with this level of detail in my amplifier design project. Not only did I double-check the operation of the circuit using a breadboard circuit, but I also learned a lot about the nuances of the transistor through developing the new model. I found that the culprit transistor was quite useful. I was able to get the proper output-swing from my amplifier, and I used the same transistor configuration in other areas of my op amp circuit. Not a waste of time for me! I gained a comfort level that is still paying me back. The product is still on the market, 14 years later.

The SPICE simulation tool is a good thing. It will help organize your thoughts and priorities. You can look for faults in terms of how you think the circuit or system will work versus reality. The best of all worlds is to have your tools point out where your mistakes are so you can initiate corrective actions to fix the circuit. The best place to find these problems is at the beginning of the design cycle, not the end. If you misunderstand the nuances of your circuit—that is, parasitics, you will find that your tools will tell you that all is well, when in fact it is not well at all.

Chapter 7

The Old Pencil and Paper Design Process

The generation of your "pencil 'n paper" circuit design is critical. Most likely, you will generate this circuit diagram using your SPICE software, but this is the stage where you take a good look at what you are trying to design. This is the time where you will develop an insider's view of your circuit. In this stage of the design, you can labor through the mathematics of your circuit or better yet, hand-wave your way through the system. During this process, you should think about device and layout parasitics. You will also define the various circuit excitation parameters in preparation for your upcoming simulation.

We've all done the mathematical calculations of our designs in our good 'ole school days. In those days, you boiled every part of the circuit down to a fundamental mathematical representation. However, the most important part of this stage is not the mathematics. In this stage, you should develop an intuition of what you think the circuit will do; then run some critical calculations in preparation for your next step, simulation. Hardware solutions (usually analog circuits) or firmware solutions (involving microprocessors, FPGAs or microcontrollers) require this process during the design phase. This technique is shown in Figure 7.4.

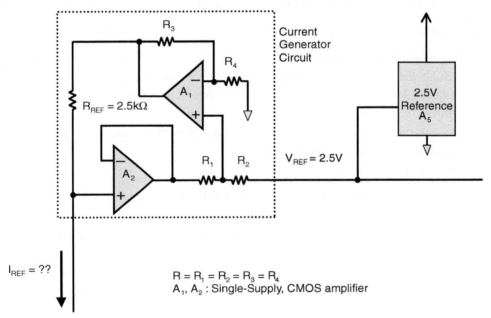

Figure 7.4: Sometimes a constant current source is required to excite a sensor in the system. This constant, floating current source contains a voltage reference, five resistors and two operational amplifiers. The fact that this current source is floating is a little tricky. This requires the pencil and paper evaluation as well as the breadboard. Conceptually, the noninverting input voltage of A_2 is independent of the 2.5 V reference (A_5).

Figure 7.4 illustrates a floating-current-source circuit design. What I want to do with this circuit is to determine what the reference current (I_{REF}) will be in relation to the reference voltage, A_5. Conceptually, the current through R_1 and R_2 is equivalent. This assumes that the input current at the noninverting input of A_1 is zero amps. If A_1 is a CMOS-input amplifier, this is a pretty good assumption. The current through R_1 and R_2 can sink into A_5 or source from A_5. The voltage at the output of A_2 can be higher or lower than the reference voltage of 2.5 V. This is actually a good thing, because we did want a floating current source.

The voltage drop across the inputs of A_1 is zero volts. If you are going to be exact, the voltage drop across these inputs is equal to the offset voltage of the amplifier, but we are going to let that go for now. The output voltage of A_1 will be at least twice as high (if not higher) as the input voltage of that same amplifier. Since the reference for R_4 is ground, the voltage of the output of A_1 will always be equal to or greater than the voltage reference.

Since this is a floating supply, the best way to get a feel for the circuit is to assign an arbitrary voltage to a node and then work out the rest of the circuit. For example, if we assume the voltage at the noninverting input of A_2 is equal to 0.5 V, the voltage at the noninverting input of A_1 is equal to 1.5 V. Given this condition, the voltage at the output of A_1 is 3.0 V. Therefore, the voltage drop across the reference resistor, R_{REF}, is 2.5 V. If R_{REF} is equal to 2.5 kΩ, the constant current source will be 1 mA.

That assumption took us a long way. It appears the impedance that I_{REF} flows through determines the voltage at the noninverting input of A_1. But, let's not be too hasty. As an exercise for you, assume that the voltage at the noninverting input of A_2 is equal to 3 V. You will find that the voltage at the output of A_1 is equal to 5.5 V. You may notice that if you have a power supply voltage of 5 V, this operation point is not possible. But, that is okay. Now we know most of the basics of this circuit. We also know the limits of the value of the resistor, R_{LOAD} in Figure 7.5.

Figure 7.5 contains a summary of the calculations for this circuit. If you follow the logic in the formulas in this figure, the voltage at the output of A_1 is equal to ½ the voltage at V_1. The voltage at the noninverting input of A_2 is equal to the voltage at the output of A_1. This voltage is also equal to the reference voltage of A_5 minus twice the difference between the reference voltage. The output of A_2 is also equal to twice V_1 minus the 2.5 V reference of A_5. The resistor value, R, is equal to 25 kΩ. This value ensures amplifier stability and to keep the output currents from the amplifiers relatively low.

Knowing this, you can determine the current through the reference resistor, R_{REF}. Thevenin says that this current is equal to the voltage drop across R_{REF} divided by R_{REF}. We can calculate this voltage drop by using the earlier equations to equal 2.5 V. As you work the real voltage and resistance values, you will summarize that the current through R_{REF} is equal to 1 mA.

Chapter 7

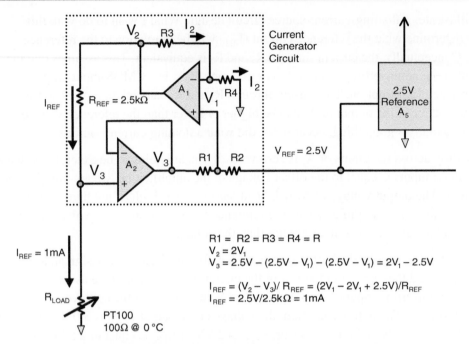

Figure 7.5: Hand-waving your way through a circuit will give you a good instinct about the circuit operation. As a final step, working through the calculations will validate your initial assumptions.

Now it is time to load the circuit. This circuit requires a low impedance load, such as a resistance temperature detector (RTD). If R_{REF} is a PT100 RTD, it is equal to 100 Ω at 0°C. In this the case, the voltage at the noninverting input of A_2 is equal to 100 mV. Consequently, the voltage at the output of A_2 is equal to 100 mV. In this application, the resistance range of the PT100 is 100 Ω @ 0°C to 254 Ω @ 400°C. At higher temperatures the output of A_1 is equal to 2.754 mV. If the power supply voltage of the amplifiers is 5 V, both amplifiers in this circuit are operating within their linear ranges.

The last steps in this portion of the process is to define the input signals, output representations of the signals, and parasitic resistances, capacitances or inductances that appear as a result of your layout of your circuit. The input signals would include transient signals in the time domain and AC signals in the frequency domain. The input signal definitions will be included at the front-end of your SPICE simulation listing. Further circuit examination will highlight the parasitic elements. For instance, the resistors in Figure 7.5 will have a parasitic capacitance (~0.5 pF) in parallel with the resistor element. Your layout may contribute additional capacitance in the ones of pico-farads because of the traces or wires that you are using. You need to determine if your PCB parasitics are an issue in your circuit. If they are, you need to quantify their values.

The stability of this circuit is another issue. Injecting a current spike into a high impedance node, such as the inverting input of A_2, will cause the circuit to ring if unstable. In the circuit in Figure 7.5, the parasitic capacitances of the resistors do not present a stability issue.

Is Your Simulation Fundamentally Valid?

Assuming you have worked through the "pencil 'n paper" design of your circuit, you are ready to simulate. The output of this first design phase should be a circuit diagram as well as the operating points throughout the circuit. Defining the operating points of the initial, DC operating points and basic operation of the circuit over time is critical. The initial DC operating point should primarily provide the node voltages, but the current magnitudes of various portions of your circuit may also be important.

Once you finish designing your SPICE model, initiate your first simulation. At the conclusion of the simulation, you should first check the validity all of the operating points in your circuit. If you miss this step, you may be looking at erroneous AC or transient simulation data. The most critical initial DC operating points are the voltages throughout the circuit. For instance, verify that your have correct power supply connections. Then check to see that all of the DC voltages in your circuit are between the power-supply voltages. If any node in the DC operating points exceed your power-supply voltage, you probably have a bad connection in your circuit net list.

Figure 7.6 contains an example of several "red flags" in the DC operating points of the Figure 7.4 circuit. This listing initially shows the simulation circuit connections. Following the .OP statement, the simulation listing calls out operational amplifier model ("ideal.mod"). Then there is a listing of the elements of the circuit, their associated device numbers and node assignments.

Everything in this listing in Figure 7.6 looks in order to this point (unless you have already found the error). All of the amplifier nodes and resistor nodes are connected. The indication that something is wrong shows up in the NODE/VOLTAGE table. This is a SPICE generated table of simulation numbers. All of the nodes are present. You won't recognize some of them because they are nodes that are internal to the two amplifier macromodels. You should immediately notice that there are negative voltages assigned to some nodes. It is a red flag that some of the negative nodes are internal in the amplifiers. This is a single-supply circuit. The supply voltages are ground (0) and 5 V (ps).

If you return to the top of Figure 7.6 you will notice there is something peculiar with the op amp, node assignments. The order of nodes versus function is:

inp1 – noninverting input,
inm1 – inverting input,
ps – positive power connect
1 – negative power connect
out1 – output

Chapter 7

```
* SHELL FOR floating current source
********************************************************************
.OP
.lib    ideal.mod
x1      inp1    inm1    ps    1    out1    mcp601
x2      inp2    out2    ps    1    out2    mcp601
* reistorsin circuit
r1              out2    inp1          25k
r2              inp1    2             25k
r3              out1    inm1    25k
r4              inm1    ns            25k
rref    out1    inp2    2.5k
rsens   inp2    ns              100
vref    2               ns            2.5V
vps     ps      0       5
vns     ns      0       0
.END
```

Some simulation results

```
****       SMALL SIGNAL BIAS SOLUTION       TEMPERATURE = 27.000 DEG C
********************************************************************
NODE   VOLTAGE    NODE   VOLTAGE    NODE   VOLTAGE    NODE   VOLTAGE

(   1)   -4.8814   (   2)   2.5000   ( ns)    0.0000   ( ps)    5.0000
( inm1)   1.3000   ( inp1)  1.3000   ( inp2)   .1000   ( out1)  2.6000
( out2)    .1000   ( x1.5)  4.9777   ( x1.6)  4.9777   ( x1.7)   .0753
( x2.5)   4.9777   ( x2.6)  4.9777   ( x2.7) -1.1247   (x1.23) -8.3335
(x1.33)    .0593   (x1.34)  3.5113   (x1.43)  2.7000   (x1.44)  2.5000
(x1.45)   4.2235   (x1.46) 45.5970   (x2.23) -4.0503   (x2.33)   .0593
(x2.34)   -.7718   (x2.43)   .2000   (x2.44) 4.276E-09 (x2.45)  4.2485
(x2.46)  45.5720
```

Figure 7.6: The first portion that you should inspect of any SPICE simulation is the DC operating points. You should check for appropriate voltage and currents in all of the elements of your circuit. This figure shows a portion of complete DC analysis. You will note that some of the nodes are negative values. Since this is a single-supply circuit, this is a warning that something is wrong.

That seems fine, but there is also a node called "ns", and it is a ground connect for the rest of the circuit. There is the error. The amplifier-macromodel, negative-supply nodes attaches to ground. The schematic capture tool generated this error.

This is just one example of where the SPICE simulation can go wrong. If you continue with any type of analysis, such as AC or time transients, you will always wonder why the results look bad. Even worse, you will do what I did in the beginning of my career and assume that these types of bad results are true. A worse case scenario is to not look at the DC operating points at all. It always pays to question results and challenge the outcome. In a particular instance that I can recall, I chased my tail for most of the week only to find out that one of the nodes was not properly connected. If I had examined the DC analysis results, I would have immediately seen the problem. But, that is what experience is all about, right?

Another place you may want to look for the correctness of your DC analysis (or further on in the simulation) is places where the default values give you erroneous results. The .OPTION

SPICE of Life

statement of SPICE contains these default values that affect a variety of conditions. The .OPTION statement sets all options, limits, and simulation analysis control parameters. The list of limits includes current accuracy, charge accuracy and the minimum conductance between branches, to name a few. It might be worth your time to look at this list. Usually these .OPTION statement defaults won't effect your simulation. However, an attitude that challenges the results of your SPICE simulation may bring errors into focus.

For instance, it may be critical to have the correct input-bias current values with a low-bias CMOS operational amplifier. An error in this parameter appears where your application circuit has high input impedances, such as a transimpedance amplifier or a low-pass filter. The SPICE program will insert a noiseless resistor inside components that have a discontinuity. The gate of the CMOS transistor is essentially floating, or not connected at DC. Although this node does have gate-to-source and gate-to drain capacitors, your SPICE will "view" this as an open circuit in the DC analysis. The SPICE software "fixes" this during the simulation by inserting a minimum conductance between discontinuous nodes.

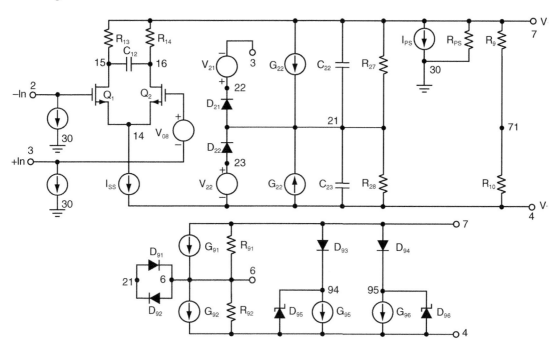

Figure 7.7: This is an example of a SPICE macromodel for a CMOS input amplifier. In this circuit, the SPICE simulation generates an input bias current error as an artifact of the SPICE constraints. Alexander and Bowers were the first to go public with this macromodel. This information appeared in the Electronic Design Magazine in 1990.

In the SPICE macromodel in Figure 7.7, the input bias current of a CMOS should be zero. Although, the ESD cells (not modeled here) because the "real" amplifier input-bias current, the CMOS transistors will not generate any current in or out of the gates of Q_1 and Q_2.

Chapter 7

In SPICE, the input or output current of the gates of these transistors is dependent on the voltage that appears across the gate-to-drain and gate-to-source nodes. This additional current is the SPICE default value, GMIN. The default value of GMIN is 1×10^{12}S (S = Siemens = $1/\Omega$). If inverted, this is equal to 10^{12} Ω. At first glance, this may not seem to be a problem. However, a voltage across that impedance will cause several pico-amperes of error. A transimpedance amplifier (see Figure 7.8) is a circuit where this error will manifest itself as an output voltage error. To solve this problem, you can change the default value of GMIN or insert voltage-dependent-current-sources from the gate to ground of Q_1 and Q_2. As a note, when you change the default values through the .OPTIONS statement, your changes will apply to the entire circuit simulation. Use this strategy with care. I prefer to make the changes to these defaults more local, which gives credence to the insertion of the voltage-dependent-current-sources over changing GMIN.

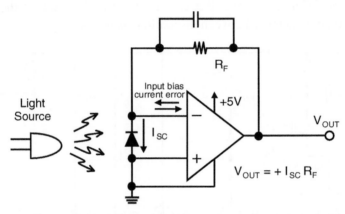

Figure 7.8: An amplifier with a low-input bias current in a transient-impedance amplifier circuit, like this one, is critical if you want to preserve reasonable accuracy. For this reason, CMOS or FET input amplifiers are preferred. If you try to simulate this circuit without the proper input bias current values, you will see an output voltage error at V_{OUT}.

Figure 7.8 highlights this problem. In Figure 7.8, impinging light generates current from the photodiode. The current then flows through R_F creating a voltage change at V_{OUT}. The light source to the photodiode generates a low-level, full-scale current of several nano-amperes. If the light is not at full-scale, generating a lower current, the amplifier input bias current errors could cause voltage errors throughout the circuit.

Your DC analysis is the most important part of the validation of your SPICE simulation. If you take the time to meticulously perform this task you will have the confidence that the rest of you simulation has a good chance of being accurate (or as accurate as the simulation can be).

You should always challenge the validity of your SPICE simulation. If you know what to expect from your simulation, you can perform these challenges. If you plan to not evaluate your circuit and just "wait and see" what your SPICE simulation produces, there is a good chance that you will either chase your tail or go back to the pencil and paper evaluation. Either way, you will have wasted valuable time.

At this point, you may have noticed a small, nagging skepticism about SPICE simulations. Well, you are right. The simulation is only as good as your imaginary SPICE circuit. The fact that you place all of the components at the proper location in your circuit diagram and you are using the manufacturers approved macromodels may give you a false sense of security! Your simulations are only as good as your models. So, how are these models defined?

Macromodels: What Can They Do?

The concept of macromodels first came about in the 1970s. ("Macromodeling of Integrated Circuit Operational Amplifiers," Graeme R. Boyle, Barry M. Cohn, Donald O. Pederson, James E. Solomon, IEEE Journal of Solid-state Circuits, Vol. SC-9, No. 6, December 1974, pp. 353–363.) These types of SPICE models provide a tool that reduces the system designer's SPICE simulation time and convergence errors. It allows the designer to focus their efforts at a higher level of simulation. However, system designers, where there are many devices in the circuit find macromodels very useful. Macromodels allow the SPICE user to simulate results successfully, in a timely fashion.

The engineers that develop IC semiconductors and use SPICE tools only use the macromodel during their circuit development to get proof of concept. They require more transistor-level detail inside their SPICE simulation when the IC designer looks at the details of their integrated circuits. As the amplifier design process progresses to completion, the macromodel simplifies the transistor-level design too much for the IC designer. Contrary to popular belief, the IC designer's models are also subject to discrepant behavior. There is no such thing as a 100% accurate SPICE model, whether it is a behavioral model, macromodel or transistor-level model.

A macromodel is actually a simple thing. On occasion, I have designed a few macromodels for existing ICs. The macromodel treats the device that you are trying to model like a black box. There are three general classes of simulation models. They are the behavioral model, the macromodel and the transistor-level model. The complexity of each of these types of models increases in the order of this list.

The behavioral model is closest to representing a "black box" with little or no relation to the actual device, except that it tries to emulate the real thing. The macromodel provides a circuit with more complexity. This type of model usually has the actual transistors on the input and output nodes. With this level of complexity, the model emulates the actual device more closely, but not completely.

In the interior of the macromodel, there are variety of dependent sources and independent sources. The most common dependent sources that are used are the linear voltage-controlled voltage source (VCVS), voltage-controlled current source (VCCS), current-controlled voltage

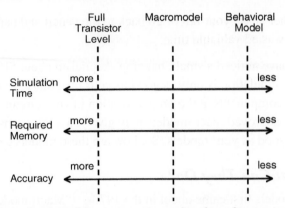

Figure 7.9: Transistor-level models are more complex than the macromodel or the behavioral model. Therefore, the transistor-level model will require more time to simulate and more computer memory. The increase in these requirements can make it difficult to complete a simulation, particularly if there are several transistor-level models in the simulation circuit. However, if you are able to tolerate longer simulation time, the transistor-level will produce results that are more accurate. In all three cases, the models are quite accurate for their intended application, and work poorly for other applications.

source (CCVS), and the current-controlled current source (CCCS). Additionally, inside the macromodel there can be nonlinear dependent sources. The two nonlinear dependent sources utilize a polynomial function in their definition. The macromodel's author defines the coefficients of this polynomial. All of these sources attempt to emulate the actual performance of the device.

The transistor-level SPICE model includes all of the transistors, resistors and inductors of the actual device. Each of the elements has a SPICE model definition. Within each of these model definitions, the user can adjust various variables. For instance, the MOSFET SPICE model has 24 variables. With these variables, the SPICE user can adjust parameters such as lateral diffusion length, lateral diffusion width, zero-bias threshold voltage, transconductance coefficient and so forth.

The transistor-level SPICE representation of the circuit is more complex than the behavioral model or the macromodel. Each of these levels has their place in your simulation strategy. The transistor SPICE model has more complexity and provides accuracy particularly for the IC designs. The transistor-level model details how the transistors in the circuit interact. This level of detail is not appropriate for the systems level designs. This type of design requires listings that can simulate several devices at one time. Under these conditions, the transistor-level models will be slower and less accurate. This is due to the increase in the node count of the entire simulation circuit.

Although the elements of a macromodel are equally complex to the transistor-level model, there are fewer elements in the total model listing. The overall complexity or sophistication of the macromodel is less than the transistor-level model. The macromodel simulates a list of specific parameters and no more. Many vendors will tell you what those parameters are in their SPICE macromodel listing. An example list of the parameters modeled for an operational amplifier would be: input voltage offset, DC PSRR, DC CMRR, input impedance, input bias current, open-loop gain, voltage ranges and supply current (typical performance at room temperature, 25°C).

A system-level simulation may include hundreds of building blocks. It may also be impossible to simulate in SPICE at the transistor or macromodel level. Even if these two levels of models do manage to converge and give results, the system level simulation accuracy is much greater with correctly modeled behavioral building blocks. The transistor-level and macromodel complexity causes reduced accuracy in this environment and greater simulation time.

Consequently, the complete device model will consume more computer memory during simulation and take longer. If you simulate several device models at the same time, such as five or six transistor-level operational amplifiers, it is possible that the SPICE simulation will "crash" and not be able to complete the simulation of the circuit. Another set back that you will find with the transistor-level model is availability. Device vendors are very reluctant to provide the transistor-level models to their customers, or for that fact, anyone. This is because the transistor-level model contains proprietary information about their circuit. You can imagine that if a competitor got their hands on this type of model, the second source design work would be reduced significantly. This would make it easy for the competition to reverse-engineer a part and to quickly start stealing market share. More importantly, transistor-level models are less accurate for board and system level designs.

Figure 7.10 shows an example of an operational amplifier macromodel.

The circuit in Figure 7.10 has some limitations. You will notice the ground connects in several places. Because of these ground connects, and the way they affect the macromodel's behavior, the model will not operate in a single-supply environment. Figure 7.7 shows a model that is

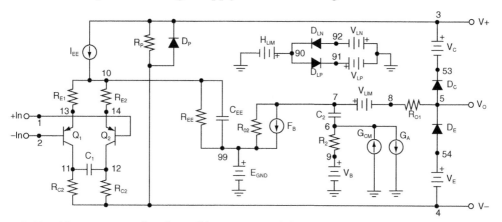

Figure 7.10: This PNP operational amplifier macromodel was originally designed in 1974 by Graeme R. Boyle, Barry M. Cohn, Donald O. Pederson, James E. Solomon. It was the first legitimate model published. Several other operational amplifier macromodel templates have been developed since then. Some SPICE vendors provide tools in their software to generate this type of macromodel.

used in single-supply or in floating-supply environments. You should notice that in the operational amplifier input stages of the models in Figure 7.7 and Figure 7.10 have transistors. The input stage is the only place in these macromodels that have any resemblance to the actual device. With transistors inserted at this point, the model emulates the unique nonlinearities of the op amp inputs. If you go beyond that stage, there is no longer a resemblance between the actual transistor-level model and these macromodels. The creator of the final model in Figure 7.7 and Figure 7.10 uses the dependent current and voltage sources to produce the real amplifier behavior. Keep in mind, the objective of the macromodel is not to copy the transistor model, but to imitate the operation of the device. In particular, the amplifier macromodel imitates the amplifier's operation in applications, and not as a stand-alone model.

With all of these issues in mind, it is easy to understand that macromodels do produce the complete performance of the actual amplifier circuit. The more simplistic models are able to simulate a limited number of the amplifier attributes. For example, in its most basic form, the macromodel illustrated in Figure 7.10 will only provide a small subset of the amplifier's attributes. This macromodel models the input bias current and input impedance input. It does not model the input rail-to-rail swing very accurately. The output characteristics that this macromodel can simulate are output current limit, output resistance and output voltage swing.

The output-voltage swing limits are set as if the amplifier is in a comparator configuration. This macromodel will not assist in demonstrating the nonlinear behavior of the output stage as the output gets close the rails. The macromodel's attributes in the AC domain include gain versus frequency, phase versus frequency and a symmetrical slew rate. This macromodel will not reproduce a real amplifier's asymmetrical (low to high, high to low) slew rate. Finally, the macromodel accurately reflects the DC quiescent current in a simulation. If the output of the amplifier is loaded and exercised, the current required will be pulled from ground and not from the power supplies. All of these performance attributes are only good at room temperature, or 25°C. Enhancements that are added by the vendor to this limited list of attributes are offset voltage at room temperature, input noise and input offset current.

Most vendors design their macromodels to produce typical performance attributes and they don't reflect the minimums and maximums you will find in the data sheet. If you want the macromodel to show you the amplifier's operation with minimum or maximum performance specifications, you will have proceed with caution and tweak the macromodel yourself. As a final shortcoming of this type of macromodel, these simulation attributes do not change as the real amplifier would if you were to vary the simulation temperature.

The model shown in Figure 7.10 is less flexible than the model shown in Figure 7.7. The model in Figure 7.7 overcomes quite a few limitations found in the model in Figure 7.10. For example, it is possible to use this macromodel for single-supply amplifiers. In addition, the output current in Figure 7.7 flows from the power supplies rather than the ground connect. Such attributes as power supply reduction (PSRR) and common-mode rejection (CMRR) ratios are included in this model. This model also lends itself to easily include over-temperature attributes.

The long and short of this discussion is that the capability of every macromodel is dependent on the whims of the macromodel designer. Figure 7.11 has a short list of a few amplifier macromodels that have various levels of capability.

	INPUT BIAS CURRENT	INPUT OFFSET CURRENT	OFFSET VOLTAGE	INPUT VOLTAGE NOISE	INPUT CURRENT NOISE	INPUT PROTECTION	INPUT IMPEDANCE	INPUT BIAS CURRENT CORRECTION	OUTPUT RESISTANCE	OUTPUT CURRENT LIMIT	OUTPUT FLOWING FROM POWER SUPPLIES	OUTPUT VOLTAGE SWING	QUIESCENT CURRENT	QUIESCENT CURRENT vs POWER SUPPLY	QUIESCENT CURRENT vs TEMPERATURE	GAIN vs FREQUENCY	GAIN vs TEMPERATURE	PHASE RESPONSE	CMRR vs FREQUENCY	PSRR	PSRR vs FREQUENCY	SLEW RATE	PAD PARASITICS	NO GROUND REFERENCE
Op amp A	X							X	X	X			X	X		X		X				X		
Op amp B	X			X	X		X		X	X	X	X	X		X	X		X				X		
Op amp C	X	X	X	X	X		X		X	X	X	X	X			X		X	X	X	X	X		X
Op amp D	X	X	X		X		X		X	X	X	X	X			X		X		X		X	X	X
Op amp E	X		X	X	X		X		X	X	X	X	X			X		X		X		X	X	X

Figure 7.11: The capability of the amplifier macromodel varies from author to author and vendor to vendor. The best line of defense is to find out what your macromodel can or can't do before you use it in your simulation. You can determine this by asking the vendor for that information, or by running your own tests in SPICE to determine what the macromodel's capabilities are.

Concluding Remarks

Computer-based simulations can significantly reduce the development time and therefore speed up the time-to-market of your designs. These facts alone make SPICE simulations attractive to the IC designer as well as the systems designer. With the increasing use of SPICE-based simulations, there is also a rising demand for accurate models. The expectation is that your models, or macromodels, reflect the actual performance of the component. This should be done without carrying the burden of too many circuit details. Companies in industry have responded to this need by providing macromodels for a broad range of products. The selection of SPICE macromodels from these companies ranges from op amps, difference amps, instrumentation amps, isolation amps, to analog function circuits.

Analog manufacturers will also provide other tools that will facilitate your design process. An analog filter's design tools are the most prevalent. Some of the tools allow the designer to use any manufacturers devices in the circuit while others require that the user use only their products. Another popular tool can help you with power supply design. In this case, the manufacturer controls the selection of the devices that will work in the circuits that their tool creates. They do this by knowing the particulars of their products and assisting design-in for their products.

Chapter 7 References

Introduction to Pspice Manual for Electronic Circuits Using OrCad Relase 9.1, Nilsson, Riedel, Prentice-Hall, 2000.

Inside SPICE: Overcoming the Obstacles of Circuit Simulation, Kielkowski, Ron M., McGraw Hill, 1994.

Macromodeling with SPICE, Choi, Pyung and Connelly, J. Alvin, Prentice-Hall, 1992.

"Spice Models Low-bias Op Amps Correctly," Baker, Bonnie C., *EDN Magazine*, August 20, 1992.

"Macromodeling of Integrated Circuit Operational Amplifiers," Graeme R. Boyle, Barry M. Cohn, Donald O. Pederson, James E. Solomon, IEEE Journal of Solid-State Circuits, Vol. SC-9, No. 6, December 1974, pp. 353–363.

"Designer's Guide to Spice-Compatible Op-amp Macromodels – Part 1," Alexander, Bowers, Electronic Design News, Volume 35, No. 4, February 15, 1990.

Semiconductor Device Modeling with Spice, Antognetti, Massobrio, McGraw Hill, 1980.

CHAPTER 8

Working the Analog Problem From the Digital Domain

CHAPTER

8

CHAPTER 8

Working the Analog Problem From the Digital Domain

When you move your analog problem solving from hardware to firmware, there can be a few useful processor or controller peripheral that will take you a long way. These peripherals include the pulse width modulator (PWM), comparator, a timer or two and the I/O gates. Although all of these peripherals sound like they are digital, we are going to use them to an analog advantage in our circuits.

The PWM may or may not be a part of your controller or processor arsenal. If the PWM function is an internal function, from your digital chip manufacturer, you can use it to produce fairly accurate voltage references. If you don't have a PWM on board, you can generate the signal in firmware. The accuracy of the voltage source generated by this tool is as accurate as your on-board timer and power supply voltage. The only thing this voltage reference requires is a PWM generator and an analog filter.

You will find that there is probably an integrated comparator in your controller or processor. If not, an I/O gate can take over this function for some applications where you are looking for a trigger or level indicator of an analog signal. I prefer working with a comparator because the threshold is usually more predictable than your run-of-the-mill digital I/O pins. You can design functions such as a window comparator or just a standard comparator with this peripheral.

An internal timer (or two) is absolutely necessary if you plan to implement the functions that we are going to talk about in this chapter. The timer of your controller or processor intimately connects to the device's clock. The clock of the device can be as accurate as you would like. Some controller or processor clocks are an internal R/C pair. These types of clocks are accurate only to a point. Other clocks inside controllers or processors are much more accurate than their R/C cousins. The accuracy of these clocks can be as good as 1 to 2% over temperature. If you are interested in finessing the accuracy of the clock, you will have to resort to using a resonator or crystal oscillator. Accurate clocks come in handy if you are implementing digital filters in the controller or processor and you want to reject the noise that is riding on a particular frequency. A popular choice for a rejection frequency of a digital filter is 50 Hz or 60 Hz. But clock accuracy is not a critical specification if you are designing a D/A converter. With this type of circuit, clock jitter affects the accuracy.

Chapter 8

Finally, to close this topic of converting analog-to-digital, this chapter will use passive components. In particular, we are going to use resistors and capacitors. The techniques and concepts that I am going to show you are not new. As a matter of fact, you learned about them in your first year of college. I always smile when I am able to use something from those days. When I am able to do that, it seems that simplicity is in charge.

Pulse Width Modulators (PWM) Used as a Digital-to-Analog Converter

There are a variety of functions and applications where a PWM comes in handy, but using it to generate an analog reference voltage is the most useful that I can think of (being the analog engineer that I am). In this section of the chapter, I will show you how to build a "poor-man's" digital-to-analog converter (DAC). The controller/DAC uses very few external parts, which makes the cost of this DAC very low.

Looking At This Reference in the Time Domain

It is possible to use the PWM to create an analog voltage. You will need not only the PWM function, but an analog filter to accomplish this. Figure 8.1 shows the timing diagram of a PWM module.

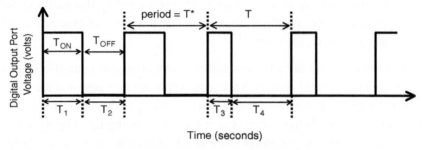

*The fundamental frequency of PWM signal = 1/T

Figure 8.1: A typical PWM waveform has a consistent period (T) with a varying ratio of on and off times. You can apply an analog filter to the output of the PWM signal to generate an analog voltage. Once you filter the PWM signal, longer T_{ON} times will generate higher voltages.

With a functioning PWM, the clock sets the fundamental frequency to the controller. After that, you can adjust the duty cycle by changing the ratio of $T_{ON}:T_{OFF}$. Figure 8.1, shows the output of a PWM as it would appear at the output of an I/O gate. When the signal is ON, the output voltage is full-scale. The actual value of this full-scale voltage is dependent on the microcontroller or microprocessor and power-supply voltage. If the power supply (V_{DD}) is equal to 5 V, the ON-voltage magnitude is ideally 5 V but in reality a few hundred millivolts below 5 V. When the signal is OFF, the output voltage is ideally 0 V, but in reality, a few hundreds of millivolts above ground. The actual value of these voltages is dependent on the specific controller or processor that you have chosen to use. The actual values of these voltages will affect the accuracy of the output analog voltage that we are going to create with an analog filter.

Working the Analog Problem From the Digital Domain

The number of divisions (K) that your clock can produce during the PWM period determines one part of the accuracy and granularity of your reference. For example, if your clock can only divide the PWM period (T) by 64, the highest granularity that you will get out of your adjustable voltage reference is 1/K = 1/64 of your full-scale range. For instance, if you are using a 5 V power supply, the smallest (ideal) change in the reference voltage is:

$\Delta V_{REF-MIN} = V_{DD}/K$

$\Delta V_{REF-MIN} = 5\ V/64$

$\Delta V_{REF-MIN} = 78\ mV$

This assumes that your PWM cycles between 0 V and 5 V. In this particular example, this is a 6-bit analog adjustable reference (or DAC), where $64 = 2^6$.

Based on the number of time divisions in the period, T, the ideal number of bits (or the resolution) of this DAC is:

DAC resolution = log (K) / log (2)   (is bits)
DAC accuracy = 6.02 N + 1.76   (in dB, where N is the number of bits)

The only problem with this voltage reference, at this point, is that the signal is still in the digital domain.

Changing This Digital Signal to Analog

An analog filter after the PWM pulse generates a DC voltage. The value of this voltage depends on the ratio of $T_{ON}:T_{OFF}$ and the power supply voltage. If the signal is ON more than it is OFF, after the filter, the output voltage will be above mid-scale (mid-scale = $V_{DD}/2$). Alternatively, if the signal is ON less than it is OFF, the output voltage will be below mid-scale. Figure 8.2 shows this relationship graphically.

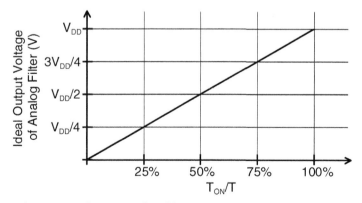

Figure 8.2: If a PWM signal is sent to an output port of the microcontroller and then through an analog low-pass filter, the output voltage of the filter with respect to V_{DD} is ratio matched to the ON time (T_{ON}/T) of the PWM signal.

Chapter 8

If you properly filter the PWM signal on the output port of the controller, it is theoretically possible to produce any analog voltage between ground and V_{DD} at the output of the filter. Wouldn't it be wonderful if the ideal theories stuck? But, this system has errors. They are the quantization error from the controller clock, the output swing of the I/O gate, the ripple rejection of the low-pass filter, and any offset errors from the low-pass, filtering amplifier. If we are in a single-supply environment, the output swing of the amplifier will never reach the rails, so you will loose a few hundred milli-volts near ground and the power supply. But let's see how close we can get to reality.

Figure 8.3 shows the suggested circuit diagram for the PWM voltage reference.

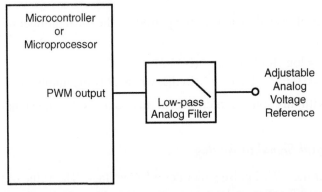

Figure 8.3: This is the hardware implementation of a PWM voltage reference using the controller or processor to generate the PWM signal. The analog low-pass filter changes the PWM signal to a DC voltage.

The most challenging part of this design is to decide on what kind of filter to use. In this discussion I am going to assume that you have read through Chapter 4, so I won't engage in another discussion on the analog filter terminology.

There are two filter specifications at the top of the list to address, corner frequency, and filter order. You can reduce the higher frequency ripple by making the corner frequency is low with respect to the fundamental frequency of the PWM. The magnitude of that reduction depends on your application requirements. If you can tolerate a ripple of 5% (which is essentially a little better than a 3-bit system), then your filter requirements will be relaxed. In contrast, if you need a reference that complements a 9-bit system, a ripple that is a little less than 0.1% is the way to go.

Defining Your Analog Low-Pass Filter for your PWM-DAC

Figure 8.1 shows the output signal of the PWM generator. Although the ratio of T_{ON}:T_{OFF} can change, the fundamental frequency never does. The fundamental frequency of the signal in Figure 8.1 is equal to 1/T. If you want to look at this signal in the frequency domain, an FFT graph is a great tool to use.

Figure 8.4 show a simplified FFT plot of the signal in Figure 8.1.

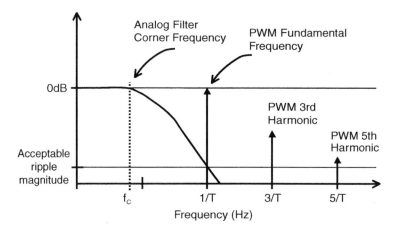

Figure 8.4: The trick to reducing the ripple at the output of the PWM generator is to apply an analog filter that averages this switching frequency to a DC result. The primary frequency generated at the output of the PWM generator is equal to 1/T. The primary frequency also has the highest magnitude of all of the harmonics. So when designing the analog low-pass filter, this fundamental frequency response dominates the calculations and results.

In Figure 8.4, you can see that the FFT plot breaks the response of the PWM into its equivalent frequencies. Refer to Appendix B for more FFT plot details with respect to the conversion of a "square" wave signal. Also in Figure 8.4, there is the frequency response of a low-pass filter. The corner frequency of this filter is lower than the PWM fundamental frequency. Your system requirements determine the analog filter, corner frequency and order. Let's look at an example with some numbers to help us understand this a little better.

Using the following parameters:

    PWM fundamental frequency, f_{PWM} = 20 kHz
    Required minimum ripple, A_{STOP} = $-$ 56 dB (8-bit accuracy to $\pm$ ¼ LSB)

The calculation of a single-pole, analog filter for this circuit is straightforward. This formula will give you your corner frequency:

$f_{C\text{-1st order filter}} = f_{PWM} / \sqrt{((10^{-A_{STOP}/20})^2 - 1)}$
$f_{C\text{-1st order filter}} = 20 \text{ kHz} / \sqrt{((10^{56/20})^2 - 1)}$
$f_{C\text{-1st order filter}} = 31 \text{ Hz}$

Chapter 8

Since this is a single-pole filter, the circuit should have an R/C pair followed by a buffer amplifier. In order to achieve a 31 Hz corner frequency, I am going to choose an arbitrary capacitor value of 1 µF. With this value the required resistor would be:

$R_{31Hz\ filter} = 1 / (2 \pi f_C C_{31\ Hz\ filter})$
$R_{31Hz\ filter} = 1 / (2 \pi\ 31\ Hz \times 1\ \mu F)$
$R_{31Hz\ filter} = 5.184\ k\Omega$ (a 5 kΩ resistor will do the trick)

If you need to back calculate the number of bits that you are actually getting out of your filter, you can use this formula:

Resolution of system in bits = $(A_{STOP} - 1.76) / 6.02$

Figure 8.5 shows the circuit diagram for this low-pass filter.

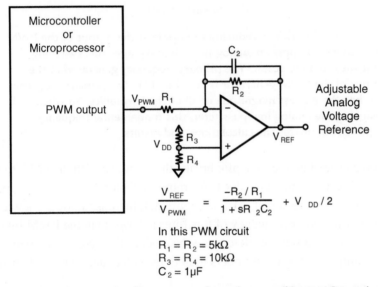

Figure 8.5: The 1st order filter uses an inverting amplifier configuration with an R/C pair in its feedback loop. This is an inverting amplifier circuit in a single-supply system. In this environment, you should reference the signal to halfway between V_{DD} and ground by using R_3 and R_4 in a voltage divider configuration.

This circuit implementation uses an amplifier with a single-pole implemented in the feedback loop (R_2 and C_2). There are two reasons to select this circuit configuration;

1. An R/C filter that is referenced to ground has considerable clipping. Erroneous signals are generated to the output,

2. Most voltage references provide a low resistance output. You can only achieve this if you use an amplifier in the circuit.

The amplifier introduces DC errors to your D/A converter. The amplifier's offset voltage goes directly to the amplifier output. This offset error will appear on every D/A converter output setting. Another words, if you expect a D/A converter output of 1.0 V and the offset voltage of your amplifier is 5 mV, the actual output of your D/A converter will be 0.990 V (assuming you are using the inverting configuration in Figure 8.5).

Besides the offset error of the amplifier, the output swing of the amplifier is limited near the positive and negative rails (in a single-supply environment the negative rail is usually ground). There are two approaches to defining output voltage swing. The output current determines how close the output can go to the rail (V_{OH} and V_{OL}); this is the most common specification. These output swing values are usually 10s of milli-volts from the rails. The definition of the amplifier's open-loop gain (A_{OL}) specification is over a smaller output swing. This smaller range is within the linear output voltage range of the amplifier. These output swing values are usually hundreds of millivolts from the rail. The output voltages of the DAC in these regions will manifest these amplifier errors.

If you need your voltage reference to remain stable under transient conditions, you may want to increase the filter corner-frequency or order. If this is the case, a higher-order filter is a good alternative, because you already have an amplifier in the circuit. Designing these filters is easy if you use the free, low-pass, filter software from various operational amplifier manufacturers. Some of the tools you may want to consider are:

Active Filter Synthesis Program	www.circuitsim.com
FilterPro Program	www.ti.com
FilterLab Program	www.microchip.com
FilterCAD Program	www.linear-tech.com
FilterWizard Program	www.analog.com

For more information about analog filters, definitions, and design, refer to Chapter 4.

Pulling the Time Domain and Frequency Domain Together

If you'll recall the time domain discussion, we found that the number of clock divisions that were possible through out the period, T, would affect the accuracy of your reference. We were able to define the accuracy in terms of bits and decibels. The discussion was in terms of time and voltage.

In contrast, when we talked about the accuracy of this reference in the analog domain we quantified the accuracy of the system in terms of decibels (dB). We did this during the examination of the response of the PWM and filter in the frequency domain. We then were able to connect the desired decibels to bits.

Assuming an ideal low-pass filter:

$\Delta V_{REF-MIN} = V_{DD} / K$ (K is number of time division in T)

DAC resolution = log (K) / log (2)   (in bits)

DAC accuracy = 6.02 N + 1.76   (in dB, where N is the number of bits)

Using a low-pass filter, assuming K is infinite:

Resolution of system = $(A_{STOP} - 1.76) / 6.02$ (in bits)

With these design equations, you can design a D/A converter that has a relatively slow output. The frequency limiting factors in this design are the clock speed of your controller's fundamental PWM signal and the cut-off frequency of the analog low-pass filter. If you are interested in improving the frequency response of this system, you can use a faster clock without compromising the resolution or you can use fewer clocks in the PWM period. This will reduce the resolution. On the analog side of this discussion, you can use a higher order filter. If you use this option in your design, the Bessel approximation type will have the best settling-time, given the input is a pseudo-square wave.

Using the Comparator for Analog Conversions

A comparator is the most common A/D converter that you will find in your processor or controller. A comparator takes an analog voltage, analyzes its value, and determines whether or not it is above or below a reference. This function is the building block of many high-level A/D converters. Converters that use comparators include the SAR, sigma-delta, voltage-to-frequency, and dual slope A/D converters, just to name a few. Later in this chapter (under the "Using the Timer and Comparator to Build a Sigma-Delta A/D Converter" section), I will use a comparator inside a modulator feedback loop to implement a 1st order sigma-delta A/D converter function.

Some of the comparator specifications of interest are the input range, input offset voltage, and input hysteresis. The diagram that I am going to refer to during this comparator specification discussion is in Figure 8.6.

Input Range of a Comparator (V_{IN+} and V_{IN-})

In a signal supply circuit, the input range of a comparator can range from one rail (ground) to the other (V_{DD}) and be able to have a reference through that entire range. But you should be aware that this specification is not always rail-to-rail. Like the operational amplifier, some comparators have one internal differential pair when others have two differential pairs in parallel. This dual differential pair will have rail-to-rail input operation. If you want more details about these types of input stages, refer to Chapter 5.

Working the Analog Problem From the Digital Domain

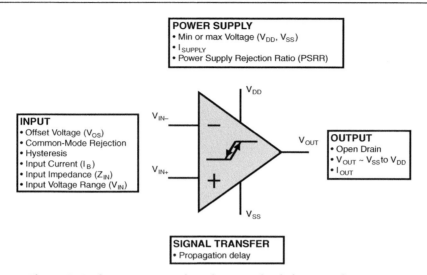

Figure 8.6: The comparator has the same basic input and output pins as the operational amplifier. There are several differences between the comparator and operational amplifier. The differences that the comparator has include a hysteresis in the input stage, propagation delay, and a digital output stage.

Input Hysteresis

A comparator must have a hysteresis on the input stage. This hysteresis will prevent the "chattering" effect on the output of the comparator when there is noise in the input signals. If you have this kind of noise is your system, Figure 8.7 shows a circuit that will extend the hysteresis quantity.

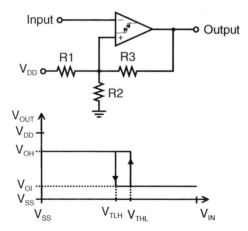

Figure 8.7: Using a feedback resistor, (R3) can enlarge the input hysteresis of a comparator, from the non-input to the output terminals. You use R1 and R2 to set a voltage reference for the comparator.

Chapter 8

You can program the trip points of the hysteresis. With V_{DD}, V_{OH} and V_{OL} known, you can calculate the resistor values with:

$$V_{AVG} = V_{TLH} + (V_{THL} - V_{TLH})/2$$
$$V_{AVG} = V_{DD} * R_2 / (R_1 + R_2)$$
$$R_{EQ} = R_1 * R_2 / (R_1 + R_2)$$
$$F_{DR} = (V_{THL} - V_{TLH}) / V_{DD}$$
$$R_3 = R_{EQ} [(1/DR - 1]$$

where,

V_{AVG} is the average between the two threshold voltages, V_{TLH}, and V_{THL};

V_{TLH} is the threshold voltage when the input signal travels from low to high. This will cause an output from high to low on the output of the comparator;

V_{THL} is the threshold voltage when the input signal travels from high to low. This will cause a transition low to high on the output of the comparator;

R_{EQ} is the parallel equivalent of resistors R1 and R2. This relationship sets up the reference voltage for the comparator circuit;

F_{DR} is the feedback divider ratio around the comparator.

Window Comparator

A window comparator comes in handy when you are monitoring an analog voltage and you would like to know when that voltage extends outside of a predetermined range. This type of circuit can provide an interrupt for your controller as the signal moves above or below a safety limit. You will design this circuit to identify instances where the input signal violates your limits, while ignoring minor fluctuations inside your safety range.

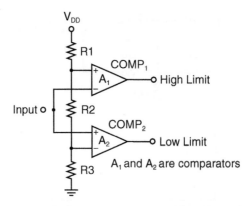

Figure 8.8: You can use a window comparator to identify analog signals that have exceeded a high or a low limit. You can accomplish this with two comparators, and by setting the threshold voltage with a voltage divider across the power supply. In this circuit, you will not require rail-to-rail input comparators.

Working the Analog Problem From the Digital Domain

The window comparator in Figure 8.8 requires two comparators and three resistors. The three resistors form a voltage divider for the references to the comparator across the supply voltage. If you want a more accurate system, you should use a precision voltage reference at the top of the resistor ladder instead of the power supply.

The outputs of $COMP_1$ and $COMP_2$ respond to the analog signal on the input of this circuit. In the event that the analog input signal goes higher than the voltage at the noninverting input of $COMP_1$, the output of $COMP_1$ will go low. This is because you are putting the analog input signal into the inverting input of the comparator. During the time where $COMP_1$ is responding to this high input voltage, $COMP_2$ remains at a logic high. Conversely, when the analog input signal goes lower than the voltage at the non-inverting input of $COMP_2$, the output of $COMP_2$ will go low. The analog "safe" zone, where both comparator outputs are high is set with the resistor R_2.

R_1, R_2 and R_3 form a voltage divider that sets the high and low threshold voltages. The design equations for this circuit are:

$$V_{TH} = V_{DD} \times (R_2 + R_3) / (R_1 + R_2 + R_3)$$
$$V_{TL} = V_{DD} \times R_3 / (R_1 + R_2 + R_3)$$

where,

V_{TH} is the high limit of the window comparator;

V_{TL} is the low limit of the window comparator.

Combining the Comparator with a Timer

Comparators alone can convert an analog signal to a digital output. This may seem to be a primitive, limited function, but when you start to combine the comparator with other peripherals in the controller or processor, its capabilities multiply. For instance, you can combine a comparator with a timer (and a few external components). With a little effort, this combination suddenly becomes a fairly good A/D converter.

Figure 8.9 illustrates a good example of using a comparator and timer to implement an A/D converter function.

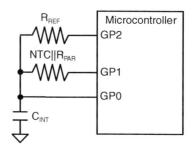

Figure 8.9: You can measure a resistor value with the controller's internal timer as long as you know the value of the reference resistor, R_{REF}.

Chapter 8

You can perform a conversion in the following fashion:

1. Set GP1 and GP2 to a high impedance inputs.
2. Set GP0 to a low impedance, low output to discharge C_{INT}.
3. Set GP0 to as a high impedance input and GP1 to a low impedance high output, start your counter. Since GP1 is high, the capacitor will start to charge.
4. When GP0 changes to 1, log the number of counts into your t_{NTC} register (t_1, Figure 8.10).
5. Set GP1 and GP2 to inputs again.
6. Set GP0 to a low output to discharge C_{INT} again.
7. Set GP0 to an input and GP2 to a high output, start your counter.
8. When GP0 changes to 1, log the number of counts into your t_{REF} register (t_2, Figure 8.10).

With this configuration, you are measuring the rise time of two R/C combinations. The first R/C combination is R_{NTC} and C_{INT}. R_{NTC} can be any type of resistor, but if it is a negative temperature coefficient (NTC) thermistor, you can measure temperature inexpensively. R_{PAR} is a resistor that you select to put in parallel with R_{NTC}. This resistor linearizes the response of the NTC thermistor, making it easier to measure the temperature range of interest.

You use a second R/C combination as your control or reference circuit. You will know the value of R_{REF} with this resistor combination. The most stable type of resistor to choose for this element is a standard wirewound, 1%, resistor. Finally, you should use a film polypropylene capacitor or any type of capacitor with low dielectric absorption for C_{INT}.

Figure 8.10 shows the time response of the algorithm, 1 through 8.

The transfer function of this system is:

$$R_{NTC} = R_{REF} \times t_{NTC} / t_{REF}$$

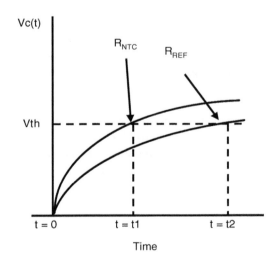

Figure 8.10: When you pull the gate that is connected to R_{NTC} or R_{REF} high after it has been in a high impedance setting, the capacitor, C_{INT}, starts to charge. The controller monitors this rise time until the voltage on top of C_{INT} equals the threshold voltage, V_{TH}. You will log the amount of time that it takes to charge the capacitor as t1.

With this circuit, if V_{TH} is ratiometric with V_{DD}, the conversion time is independent of the power supply voltage, V_{DD}. V_{TH} is set by the input of an I/O port or by a comparator. Either type of inputs will respond reliably. You will note from the transfer function of the calculation is independent of the capacitor and the clock frequency of your controller. In this system, the resolution is dependent of the R/C time constants and the granularity of your timer. You determine the accuracy with R_{REF}.

Using the Timer and Comparator to Build a Sigma-Delta A/D Converter

This portion of this chapter describes how to implement a sigma-delta, A/D converter function using a microcontroller. Although many microcontrollers do not have a built-in A/D converter, you can use the comparator function, internal voltage reference, and timers to digitize an analog signal.

Some of the standard controllers have a comparator module, consisting of two comparators. You can use a controller's internal voltage reference source with the comparators to establish thresholds. If this internal voltage reference is not available, you can use an external reference. By combining these elements, you can design a 1st order modulator and 1st order filter. This combination emulates the function of an analog-to-digital, sigma-delta conversion.

You can quickly implement this method of conversion in firmware, with very few additional external components. Therefore, the cost of hardware implementation is minimal, particularly for such a high-resolution converter solution. The input range is very flexible and adjusted with external resistors. Although this method is not particularly strong in terms of DC accuracy, it is well suited for ratiometric applications.

Sigma-Delta Theory

The function of the classical sigma-delta analog-to-digital converter uses two circuit segments; a modulator and a digital filter. The modulator section acquires an input signal as shown in Figure 8.11. This is the same type of modulator that was discussed in Chapter 2 (Figure 2.18).

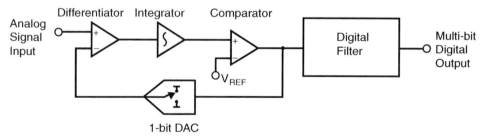

Figure 8.11: This is the modulator sigma-delta A/D converter. In the stand-alone sigma-delta A/D converter, the modulator, low-frequency noise moves out into higher frequencies. This 1st order modulator can achieve accurate conversions up to about 10-bits, without additional digital filtering.

Chapter 8

In the discussion in Chapter 2, I mentioned that this block diagram was simply a concept for discussion. Stand-alone sigma-delta A/D converters have more than one modulator segment and they also accept differential input signals. The modulator block in Figure 8.11 is a 1st order modulator that only accepts single-ended inputs.

In this block diagram, the circuit subtracts the input signal from a digital-to-analog (D/A) converter signal in the negative feedback loop. Then the differentiated signal passes through an integrator and finally to one of the two inputs of a comparator. The comparator acts like a one-bit quantizer. The comparator sends its signal back to the differentiator by way of a one-bit D/A converter. Additionally, the output of the comparator passes through a digital filter. The complexity of this digital filter is up to you, the controller programmer. With time, the output of the digital filter provides a multibit conversion result. As mention before, this fundamental circuit concept generates a large variety of the converters that provide high resolution, relatively inexpensively.

The next logical step for this type of A/D converter is to move it into the controller. A basic controller is not able to execute this type of function, however, a few additional peripherals make it possible. Figure 8.12 shows the circuit diagram for this type of microcontroller implementation. The circuit in Figure 8.12 transforms the theoretical concept in Figure 8.11 to reality.

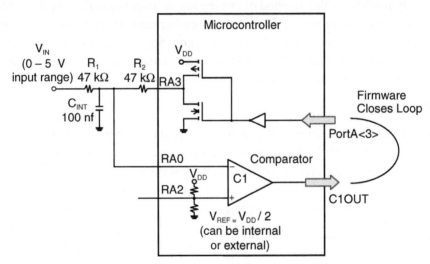

Figure 8.12: If a microcontroller has an internal comparator and timer, you can build a sigma-delta converter with two additional external resistors and one capacitor. In this configuration, a low-pass filter is a by-product of the external resistors and capacitor input network.

Figure 8.12 shows a circuit that has the integrator function in the sigma-delta block diagram of Figure 8.11. The external capacitor, C_{INT}, implements this integrator function. The absolute accuracy of this external capacitor is not critical, only its stability from integration to integration, which occurs in a relatively short period of time. When RA3 of the microcontroller is set high, the voltage at RA0 increases in magnitude. This occurs until the output of the comparator (C1OUT) is triggered low. At this point, the driver to the RA3 output is switched from high to low. Once this has occurred, the voltage at the input to the comparator (RA0) decreases. This occurs until the comparator is tripped high. At this point, RA3 is set high and the cycle repeats. While the modulator section of this circuit is cycling, two counters keep track of the time and of the number of ones versus zeros that occur at the output of the comparator.

The comparator is part of the controller, as well as its voltage reference. You can implement the one-bit D/A converter in firmware by driving RA3 in accordance with the output of the comparator (CMCON<6>, PIC16C623 from Microchip Technology). The firmware drives the D/A converter output at RA3. Two counters implement a 1st order, digital filter (also known as a averaging filter).

The Controller Implementation

With the circuit in Figure 8.12, it is possible to conceptualize the sigma-delta function. Figure 8.13 summarizes the controller implementation of this circuit in the flow chart.

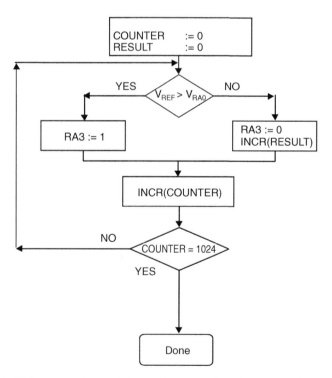

Figure 8.13: This is the sigma-delta A/D conversion flow chart, from the circuit shown in Figure 8.12. You should ensure that the cycle time through the flow chart is constant. This code runs until a conversion is complete.

Chapter 8

Normally the output of the comparator is connected to RA3. This keeps the voltage at RA0 equal to the reference-voltage of the comparator, in preparation for the next conversion. At the start of this flow chart, the result and counter variables are cleared.

You should check the comparator at the beginning of each loop. If the voltage on the capacitor is less than the input voltage, RA3 is set high, which will put charge into the capacitor, raising the voltage. If the voltage on the capacitor is greater than the input voltage, RA3 will be set low, taking charge out of the capacitor lowering the capacitor voltage and the result register is incremented.

This continues as long as necessary to get the required resolution. For ten bits of resolution, 2^{10} (1024) laps through the loop are required.

You would take each integration result at a regular time interval. If you assume that the time interval of a conversion is 20 msec, you can easily calculate the conversion time versus bits. Figure 8.14 shows this relationship graphically.

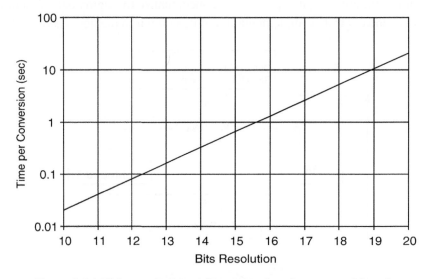

Figure 8.14: This graph shows the conversion time versus bits of resolution, assuming a 20 msec integration time while using the circuit shown in Figure 8.12.

For instance, a 10-bit conversion would require 2^{10} or 1024 samples. If the microcontroller conversion loop is 20 µs, one complete conversion would take a little more than 20 ms. Figure 8.15 shows the room temperature test data for the circuit that uses a PIC16C623 from Microchip.

Figure 8.15 shows the plot of the voltage input versus the output code on the left axis and the output error on the right axis. This data represents the results of 1024 laps through the flow chart in Figure 8.13. The expected resolution of this configuration is 10-bits. The maximum

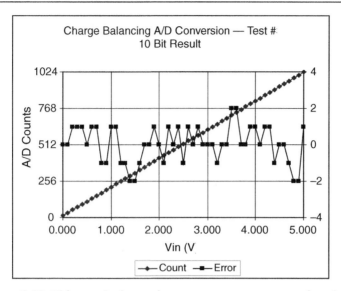

Figure 8.15: This graph shows the room temperature test data for the circuit shown in Figure 8.12 and a PIC16C623 microcontroller from Microchip. The input voltage range is 0.003 to 4.99 V. The maximum error found in the test was ±2 counts. In this 10-bit system, the ±2 LSBs is equivalent to ±9.8 mV. The data from the A/D counts (left y-axis) was only one sample per point on the graph. Results may vary from part to part. V_{DD} = 5 V.

code error for this test was ±2 counts or 2-bits of uncertainty. Consequently, the effective number of bits of this A/D converter is 8-bits.

The A/D error was calculated assuming the codes for Vin = 0.5 V and Vin = 4.5 V are ideal. The test conditions are room temperature with one microcontroller. These results may vary from part to part.

Error Analysis of this Sigma-delta A/D Converter Implemented with a Controller

This low cost sigma-delta converter provides a good solution for ratiometric applications where having the absolute results is not critical. Additionally, a 1st order, digital, FIR filter replaces the analog gain function. This FIR filter is a simple accumulator. In this example, V_{DD} is 5 V and the reference voltage is $\sim V_{DD}/2$. The resistors are 47 kΩ. This value of resistance minimizes the leakage errors across the resistors versus the RDS_{ON} error of the output pin, RA3. The capacitor value is equal to 100 nF.

RDS ON Error

This error comes from the drain-source resistance of the output FETs on the output pin, RA3. At room temperature, this resistance error is small and is typically less than 100 Ω. Compared to R_2, RDS_{ON} introduces about 0.2% gain error. You can compensate for this error

Chapter 8

by increasing the resistor, R_1 by approximately 100 Ω. Additionally, the value of the RDS_{ON} resistance will probably increase with rising temperature. Refer to these specifications in the product data sheet of your controller or processor.

RA0 Port Leakage Current

A typical specification for leakage current of the PIC16C623 is 1 nA at room temperature and 0.5 mA (max) over temperature. The leakage current from the port at RA0 causes a voltage drop across the parallel combination of R_1 and R_2. With these two resistors equaling 47 kΩ, the error caused by this leakage current is ~11 mV. This is also close to a 0.2% error. At room temperature, this error is negligible. Leakage current does increase with temperature. Refer to these specifications in the product data sheet of your controller or processor.

Nonsymmetrical Output Port (RA3)

When the output port is high, the FET resistance is dependent on the p-channel, on-resistance. When the output port is low, the FET resistance is dependent on the n-channel, on-resistance. The p-channel, on-resistance is usually greater than the on-resistance of the n-channel FET. As a consequence, there is an additional offset contribution of 5.5 mV at room and over temperature. Refer to these specifications in the product data sheet of your controller or processor.

Voltage Reference

The internal voltage reference to the comparator is a simple internal voltage divider. If this is the case, the absolute value of this voltage is dependent on internal resistor matching and power-supply voltage. Assuming the power supply is an accurate 5 V, the voltage error of this reference, part to part is significant. However, once you remove the initial error of the internal voltage reference through calibration, it is ratiometric to the power supply. This is the biggest error in the circuit, but easily reduced with an external voltage reference. The design equations for this circuit are:

$$V_{IN(CM)} = V_{RA0}$$
$$V_{IN(P\,TO\,P)} = V_{RA3(P\,TO\,P)} (R_1/R_2)$$

where,

$V_{IN(CM)}$ is equal to $(V_{IN\,(MAX)} - V_{IN\,(MIN)})/2 + V_{IN\,(MIN)}$
V_{RA0} is the voltage applied to the comparator's inverting input
$V_{IN\,(P\,TO\,P)}$ is equal to $(V_{IN(MAX)} - V_{IN(MIN)})$
$V_{RA3\,(P\,TO\,P)}$ is equal to $V_{RA3(MAX)} - V_{RA3(MIN)}$

Refer to these specifications in the product data sheet of your controller or processor.

Other Input Ranges

Figure 8.12 shows a configuration that uses a 0 to 5 V input range. The resistor network (R_1 and R_2) and the reference voltage to the noninverting input of the comparator determine

Working the Analog Problem From the Digital Domain

the input range for this circuit. If the ratio of R_1 and R_2 is changed, the input range can be increased or decreased in accordance with the relationship between R_1 and R_2. You can implement further adjustments by adding an additional resistor to this input structure that is biased to ground or the power supply.

Input Range of 2 V to 3 V

You can increase or decrease the input range of this converter by adjusting the ratio of R_1 and R_2. In Figure 8.16 these resistors reduces the input range from ±2.5 V. In Figure 8.12 the range is ±500 mV. In both cases, the input range is centered around the comparator, reference voltage, 2.5 V. This type of input range is best suited for sensors with smaller output voltage ranges, such as the buffered output of a pressure sensor or load cell.

The resistors are determined by comparing the desired input range to the voltage range of RA3. Assuming that the reference voltage in this problem is 2.5 V, the input range changes ±500 mV and the voltage at RA3 changes by ±2.5 V. The ratio of these two voltage ranges is 5:1. Consequently, during one integration period, the difference between the current through R_2 and R_1 must always be less than zero. In this manner, the RA3 gate will be capable of driving the capacitor, C_{INT}, past the reference voltage applied to the non-inverting input of the comparator. Figure 8.16. shows this circuit.

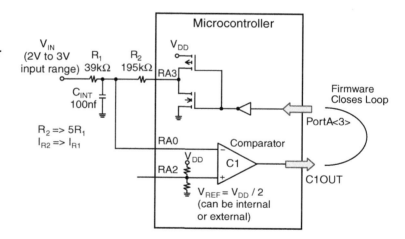

Figure 8.16: This is the configuration for the microcontroller of a sigma-delta conversion with a ±500 mV range centered around 2.5 V.

The design equations for this circuit are:

$$V_{IN(CM)} = V_{RA0}(1 + R_1/R_3)$$
$$V_{IN(P\ TO\ P)} = V_{RA3(P\ TO\ P)}(R_1/R_2)$$

where,

$V_{IN(CM)}$ is equal to $(V_{IN(MAX)} - V_{IN(MIN)})/2 + V_{IN(MIN)}$
V_{RA0} is the voltage applied to the comparator's inverting input
$V_{IN(P\ TO\ P)}$ is equal to $(V_{IN(MAX)} - V_{IN(MIN)})$
$V_{RA3(P\ TO\ P)}$ is equal to $V_{RA3(MAX)} - V_{RA3(MIN)}$

Chapter 8

Input Range of 10 V to 15 V

You can apply an offset adjustment by adding an additional resistor to the input structure of the A/D converter. In Figure 8.17, R_1 and R_2 are equal and configured to allow for an input range of ±2.5 V as shown in Figure 8.12. The addition of R_3, which is referenced to ground, provides a level shift to the input range of 10 V.

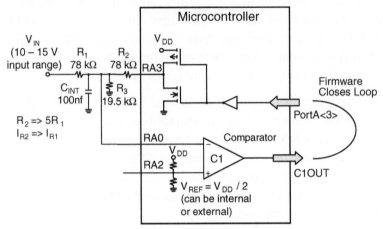

Figure 8.17: Configuration of the microcontroller for a sigma-delta conversion with a 10 V to 15 V range centered around 12.5 V.

With this circuit configuration, a 5 V (full-scale) current through R_1 is equal to V_{REF}/R_1. If R_3 draws the same current to ground, the integrating capacitor will not be charged. In this manner, a 2.5 V offset is implemented with $R_3 = R_1$. To achieve a 10 V offset, R_3 must be equal to $4*R_1$, per Figure 8.17.

The design equations for this circuit are:

$V_{IN(CM)} = V_{RA0} (1 + R_1/R_3)$
$V_{IN(P\ TO\ P)} = V_{RA3(P\ TO\ P)} (R_1/R_2)$
where,

$V_{IN(CM)}$ is equal to $(V_{IN\ (MAX)} - V_{IN\ (MIN)})/2 + V_{IN\ (MIN)}$
V_{RA0} is the voltage applied to the comparator's inverting input
$V_{IN\ (P\ TO\ P)}$ is equal to $(V_{IN(MAX)} - V_{IN(MIN)})$
$V_{RA3\ (P\ TO\ P)}$ is equal to $V_{RA3(MAX)} - V_{RA3(MIN)}$

Input Range of ±500 mV

The circuit in Figure 8.17 using the scaling technique discussed in the circuit shown in Figure 8.16, and the offset shift technique discussed in the circuit shown in Figure 8.16. With this circuit, the input range is ±500 mV. You achieve this by making $R_2 = 5R_1$. There is a level shift of –2.5 V to the signal-input range. You can implement this with a resistor, R_3, to the positive supply, per Figure 8.18. The magnitude of this level shift is achieved by making $R_3 = R_1$.

Working the Analog Problem From the Digital Domain

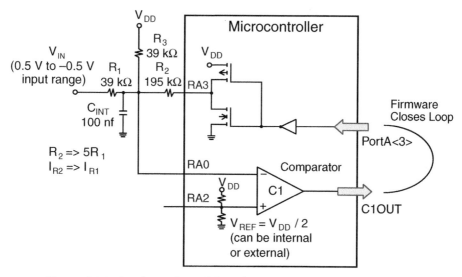

Figure 8.18: Configuration of the microcontroller for a sigma-delta conversion with a +/-500 mV range centered around ground.

The design equations for this circuit are:

$V_{IN(CM)} = V_{RA0} (1 + R_1/R_3)$
$V_{IN(P\ TO\ P)} = V_{RA3(P\ TO\ P)} (R_1/R_2)$
where,
  $V_{IN(CM)}$ is equal to $(V_{IN\ (MAX)} - V_{IN\ (MIN)})/2 + V_{IN\ (MIN)}$
  V_{RA0} is the voltage applied to the comparator's inverting input
  $V_{IN\ (P\ TO\ P)}$ is equal to $(V_{IN(MAX)} - V_{IN(MIN)})$
  $V_{RA3\ (P\ TO\ P)}$ is equal to $V_{RA3(MAX)} - V_{RA3(MIN)}$

You can use this circuit to measure the current through a shunt resistor. The main error term at room temperature is comparator offset. In systems with a known "zero-current" state, the offset can be measured and removed through calculation or removed by adding or subtracting the offset of the result counter.

Conclusion

So, you may have thought that you were using digital devices when you chose to work with a microcontroller or microprocessor. You may also think that the bridge to analog will never happen with your devices. But, in fact, the peripherals like counters, I/O ports, and comparators allow you to get much closer to analog than you might have imagined. The combinations of these tools that we discussed in this chapter, will not replace the high-precision or high-speed analog circuits. However, in your designs you need to consider how much is good enough. If you conclude that the controller can get you the analog functionality that you want, then, have at it.

Chapter 8 References

"Implementing Ohmmeter/Temperature Sensor," Cox, Doug, AN512, Microchip Technology, Inc.

"Resistance and Capacitance Meter Using a PIC16C622," Richey, Rodger, AN611, Microchip Technology, Inc.

"Analog Design in a Digital World Using Mixed Signal Controllers," Curtis, Keith, AN823, Microchip Technology, Inc.

"D/A Conversion using PWM and R2R Ladders to Generate Sine and DTMF Waveforms," Stein, Day, AN655, Microchip Technology, Inc.

"Using PWM to Generate Analog Output," Palacherla, Amar, AN538, Microchip Technology, Inc.

CHAPTER 9

Systems Where Analog and Digital Work Together

CHAPTER 9

Systems Where Analog and Digital Work Together

We've come to expect more from our battery-powered equipment. The first personal data assistant (PDA) that I purchased would only retain its battery power for the day. If I enabled the PDA calendar-alarm, the battery power would quickly go to nothing. Today, my PDA holds its battery charge for an entire week under the same conditions. Just so you know, I am comparing apples to apples. Both PDAs, from the same manufacturer, had the same battery (Li-Ion) with the same power density. So what had changed? Simple. The electronic hardware was improved and the software, power-management techniques were refined. Battery improvements were secondary.

This is a great example where system sophistication is improving rapidly, but as fate would have it, the battery-power requirements are increasing while battery chemistries are not keeping up with the pace. You can accomplish this increase in equipment functionality only if a savvy, firmware/software engineer understands the tools available in the microcontroller and the hardware designers understand the efficiency of the solutions available on the market today.

This chapter starts by reviewing battery technologies. From this short discussion, you will have an idea on how to select the correct battery for your application. One of the primary challenges you will face when you put together a battery-powered system is determining the power-management strategy. Beyond the selection of the battery, where chemistry, charging methodologies, protection and fuel gauges are a major concern, the selection of the power-conversion strategy has a significant effect on the dynamic performance and efficiency of the system. Following the battery chemistry section, we will evaluate the various hardware techniques that convert the battery voltage (that varies with time) with power efficient hardware.

At this point, assuming we have a good, efficient power supply for your circuit, we will discuss the microcontroller programming tricks that further reduce power in your circuit. This section will highlight clocking tricks and power down tips that can be implemented in software.

Finally, you will see an example of combining microcontroller. programming tricks with analog power supply management. This example will show you how to further reduce the power in your battery circuit.

Chapter 9

Selecting the Right Battery Chemistry for Your Application

Batteries fall into two fundamental categories: Primary cells and secondary cells. The difference between these two types of cells is their ability to recharge after use. The primary cell battery is not rechargeable. You would throw these types of batteries away after use. The more common chemistries for these types of batteries are: zinc, carbon, alkaline and lithium.

The most popular chemistry used for the primary cell is alkaline, with lithium on the rise. You probably recognize this battery chemistry. You can find this type of battery in electronic calculators, cameras, electric shavers, tape recorders and remote controllers. The attributes that these applications have in common are their requirements for larger current/battery capacity, lower self-discharge, low internal resistance and low cost/ease of replacement capability. Table 9.1 shows examples of some of the typical performance specifications for alkaline batteries.

Table 9.1: This table shows the typical alkaline battery specification. The definitions of specifications are: nominal voltage—typical operating voltage for the cell; rated capacity—amount of energy available until the cell reaches the cut-off voltage; rated cut-off voltage—the minimum operating voltage; and energy density by weight—amount of energy available in the battery based on weight.

Battery	Nominal Voltage (V)	Rated Capacity (mAh)	Rated Cut-off Voltage (V)	Energy Density by Weight (mWh/g)
D	1.5	17,000	0.8	180
C	1.5	7,800	0.8	167
AA	1.5	2,780	0.8	179
AAA	1.5	1,150	0.8	143
9V	9.0	570	4.8	114

The alkaline battery is a good "workhorse" for everyday flashlights, radios, and toys. You can double or triple the nominal voltage of 1.5 V by putting batteries in series. The rated capacity is significantly larger than secondary cells such as nickel-cadmium (NiCd), nickel metal hydride (NiMH) or lithium-ion (Li-Ion) batteries.

Secondary-battery cells are rechargeable. The more typical secondary cells include sealed lead acid, NiCd, NiMH, Li-Ion and lithium-polymer (Li-Poly). You will find the sealed lead acid battery in automobiles or applications where weight and size is a secondary consideration. If weight and size is a consideration, the NiCd, NiMH and Li-Ion batteries are the batteries of choice for portable applications. The Li-Poly and fuel cell batteries are new to the market. In this chapter, we will not discuss these two battery types.

Table 9.2 summarizes the general applications and characteristics of the secondary batteries used in portable applications.

Depending on the application, these three types of secondary-battery cells each have their own set of advantages and disadvantages. The NiCd battery was the first major rechargeable

Systems Where Analog and Digital Work Together

Table 9.2: This figure shows the typical secondary-battery cell applications and specifications. The definitions of the specifications are: energy density by weight—the ratio of available energy to its weight; operating voltage—a typical operating voltage for a fully charged cell; primary charge termination method—is used to identify a fully charged battery. $-\Delta V/dt$ uses a drop in battery cell voltage over a predetermined time. $-\Delta T/dt$ uses a reduction in battery cell temperature over a predetermined time. IMIN uses a measurement of the current into the battery during charge.

Battery Chemistry	Some Typical Applications	Energy Density by Weight (Whr/kg)	Operating Voltage	Primary Charge Termination Method
NiCd	Power tools, electronic tools	40–80	1.2	$-\Delta V/dt$
NiMH	Shavers, digital cordless phones, toys. May replace NiCd. More environmentally friendly than NiCd.	60–100	1.3	$-\Delta V/dt$ or $-\Delta T/dt$
Li-Ion	Cellular phones, notebook PC	110–130	3.6	I_{MIN} + Timer

battery on the market and is capable of supplying larger current spikes. This is particularly useful in applications such as power tools. If the battery has time to charge slowly, the electronics to support this battery is relatively inexpensive. On the other hand, the energy density of this type of battery is significantly lower than the other two types shown in Table 9.2. In addition, the NiCd battery has some environmental issues in terms of storage and shipping. Basically, it is not "green" friendly.

Another secondary-battery type is the NiMH battery. This battery followed the NiCd battery to the marketplace with improved energy density. It is also environmentally friendly. As compared to the NiCd battery, the NiMH battery does not have the ability to effectively handle large current rushes as the NiCd does, and the charging electronics are a little more sophisticated.

The more popular secondary-battery cell for portable applications is the Li-Ion cell. This battery type followed the NiCd and NiMH to the marketplace. The energy density of the Li-Ion cell today is best in class.

Taking the Battery Voltage to a Useful System Voltage

As you select your application's battery power-management chips, efficiency, performance, cost, and size are some of the considerations you might want to make. The choices for this application problem are switched power converters (SPC), charge pumps, and low dropout regulators (LDO). As far as efficiency is concerned, the SPC is overall the best of class for almost all applications. If the application can tolerate the switching electro-magnetic interference (EMI) noise, its efficiency is relatively independent of line voltage and output current. Ideally, the SPC efficiency is 100%, the charge pump is nearly comparable to the SPC solution with power efficiency in terms of line voltage, but is more efficient for a small range of load current. However, typically charge pumps have unregulated outputs. If you need to regulate the output of a charge pump, you need a LDO at the output of the charge pump. The efficiencies of the charge pump and LDO multiply. The result is a system with a lower

Chapter 9

efficiency than either the LDO or the charge pump alone. The unregulated linear regulator efficiency is approximately the ratio of V_{OUT} over V_{IN}. Given this scenario, the efficiency is dynamic and decreases linearly with the increase in line voltage. If the source voltage has wide changes, the linear regulator is the poorer choice for your power-supply design. But, the good news is that a linear regulator's efficiency is relatively independent of output-current load if the application can tolerate the power dissipation.

With all of these options, it would seem that choosing the right power system for your application would be difficult. But if you take the time, careful selection of the power-supply strategy can provide a competitive edge by providing the most efficient, compact, low-cost solution for your circuit.

Defining Power Supply Efficiency

You can implement the power-supply management circuit block diagram using discrete components, or a combination of integrated circuits and discrete components (see Figure 9.1). This power-supply circuit converts the changing source voltage (discharging battery) to a steady output voltage, while complementing the output drive requirements. In the best of cases, optimum efficiency is part of this task's objective. Typically, you will use the SPCs, charge pumps, or LDOs integrated circuits. In all cases, the integrated circuit conditions the source voltage to a different output voltage.

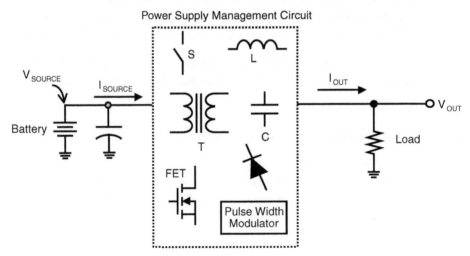

Figure 9.1: You can implement the control and reconfiguration of a power supply, such as a battery, with the three fundamental devices: SPC, charge pump or LDO. These three devices use a variety of building blocks to achieve the power-supply voltage and current transformations.

Systems Where Analog and Digital Work Together

The Efficiency of the Buck-SPC Circuit

Figure 9.2 illustrates a simplified example of a Buck-SPC circuit. With this style of converter, a simple chopping network in combination with a low-pass LC (inductor-capacitor) filter is used. In this discussion, the Buck converter is operating in a continuous inductor current mode. The input is "chopped" using a pulse width modulator (PWM) signal to the switch with resulting pulses that are averaged to create a DC-output voltage. This converter is only capable of stepping down the input voltage (from a high value to a lower value).

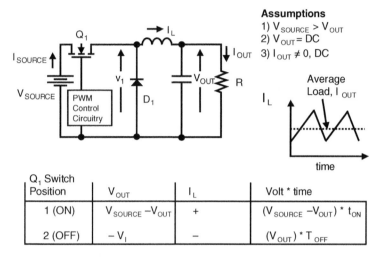

Figure 9.2: This illustration shows a Buck-SPC with the dynamic evaluation calculated for the device in the vontinuous mode. This power-supply solution is the most robust between a charge pump and LDO circuit.

Evaluation of the circuit in Figure 9.2 uses the following assumptions:

1. The input voltage is always greater than the output voltage.
2. The output voltage is essentially DC, which implies that the output filter is large enough to average this voltage, (typically <1% variation on V_{OUT}).
3. There is always a current through the inductor, which is required when keeping the converter in a continuous conduction mode.

With these assumptions, the Buck-SPC stage has two states per switching cycle. The ON state is when Q_1 is ON, and D_1 is OFF. The OFF state is when Q_1 is OFF and D_1 is ON. The duration of the ON state is equal to $(D \times t_S) = t_{ON}$. The OFF state is equal to $[(1-D) \times t_S] = t_{OFF}$, where D is the duty cycle set by the PWM control circuitry, and t_S is the switching period or 1/Fsw.

$$V_{OUT} = V_{SOURCE} \times D$$
$$I_{OUT} = I_{L(AVG)}$$

Chapter 9

Several issues have an impact on the efficiency of the Buck-SPC circuit. Four main sources usually account for most of the efficiency losses. Three of the four sources of loss are associated with the MOSFET. The first source is the gate charge-current of the MOSFET that is a result of the PWM switching action. This loss is almost independent of load. The second is the power dissipation of Vsw × Isw while the MOSFET is in the linear region during transition. You can improve this by making the edges faster, which in turn will introduce more conducted and radiated emissions noise. A third source of loss is the RDSON of the MOSFET and other resistances in the circuit. In a good design, the efficiency curve will peak at full-load or just before full-load. At that point, the switching losses are about equal to the conduction or I^2R losses. When the efficiency curve starts to roll-off, the ON resistance of the MOSFET will start to dominate. Moreover, the output diode is a major source of power loss, particularly at higher currents.

The Efficiency of the Charge Pump Circuit

Figure 9.3 shows a simplified circuit example of the charge pump, power-management system. This is a simplified circuit diagram of an inverting-charge pump. The inverting charge pump configuration inverts the input voltage, V_{IN}, to the V_{OUT} node. For example, by applying +5 V to V_{IN}, a −5 V voltage appears at V_{OUT}. The ideal switched capacitor, charge-pump converter has a two-phase oscillator, four switches (S_1 through S_4) that are synchronized by a two-phase oscillator and two capacitors (C_1 and C_2). C_1 is the 'pump' capacitor or 'flying' capacitor and C_2 is the 'reservoir' or 'output' capacitor.

The voltage inversion occurs as a result of a two-phase operation. During the first phase, switches S_2 and S_4 are open, and switches S_1 and S_3 are closed. During this phase C_1 charges

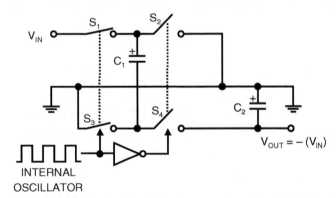

Figure 9.3: This diagram shows an ideal switched capacitor charge-pump inverter. The transfer function is $V_{OUT} = -V_{IN}$. You can immediately see that the input voltage is inverted, which is a nice feature if you are trying to generate a dual supply from a single-supply voltage. You will also note that the output "tracks" the input quite closely. Consequently, the output follows any variations on the input voltage, without regulation.

to the voltage, V_{IN}; meanwhile, C_2 supplies the load current. In the second phase, switches S_2 and S_4 are closed, and switches S_1 and S_3 are open. In this configuration, C_1 and C_2 are in parallel with the topside of C_1, which was connected to V_{IN}, is now connected to ground. The bottom side of C_1, which was connected to ground, is now connected to the bottom side of C_2. This action causes the transfer of the charge from C_1 to C_2. These connections result in a negative V_{IN} voltage at the V_{OUT} node.

The losses in this network comprise of the effective switching resistance (as it relates to the switching frequency), actual switch resistances, and the results of the effective series resistance of C_1 and C_2. Most typically, the actual switch resistances dominate the performance.

The Efficiency of the Low Dropout Regulator

You will find low dropout regulators (LDOs) in many portable applications, including cell phones, pagers and PDAs. The first LDOs to market were fabricated using bipolar technologies. These devices had and have a number of useful features. Their small size, low output noise, and precision output voltage are attributes that are well suited for battery-powered applications. However, bipolar LDO products have the disadvantage of having higher dropout voltages and exhibiting excessive ground currents as compared to the newer technologies, such as the complementary metal-oxide semiconductor (CMOS) processed devices.

Today, CMOS technologies have been developed so that this small geometry process can be designed to meet most of the bipolar LDOs' features. But beyond the added benefit of consuming less silicon real estate, the CMOS LDOs have lower drop-out voltages and they have dramatically reduced the problem of excessive ground currents with changing output loads or input voltages.

Figure 9.4 shows the fundamental topology of the bipolar LDO and CMOS LDO. The bipolar LDO block diagram in Figure 9.4a shows that this type of LDO is fundamentally constructed using a bandgap voltage reference, an operational amplifier, and output PNP pass transistor. The voltage reference is the core of this device providing voltage output accuracy through loaded conditions and over temperature. An operational amplifier buffers the voltage from this voltage reference. This high gain amplifier not only buffers the PNP pass transistor from the voltage reference, but it ensures that the LDO has good output current load regulation and input voltage ripple rejection. The third element, the PNP pass transistor, is capable of driving currents to the circuit's load through the OUTPUT terminal.

The drop-out voltage of these devices is defined as the smallest voltage between the INPUT and at the OUTPUT before the output voltage falls by 2% of its original voltage. For the PNP pass transistor bipolar LDO, the dropout voltage is typically ~0.3V. This voltage can be higher or lower, depending on the topology of the pass transistor cell. But generally is not as low as the CMOS counterpart.

Again, in Figure 9.4a, it is easy to see that the base current of the PNP pass transistor will increase when the input voltage nears the regulated output voltage. The primary reason for

Chapter 9

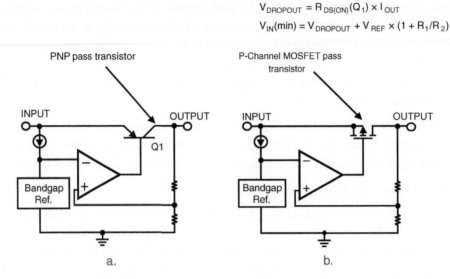

Figure 9.4: The LDO that uses a PNP pass transistor (a) sinks current from the base of the transistor to ground. Under similar conditions, the LDO that uses a P-Channel MOSFET (b) does not conduct current from the MOSFET gate to ground.

this additional, unwanted current is a result of increases in the base current of the internal bipolar transistor when the transistor goes into saturation. To worsen this circumstance, this occurs when the battery is at a weakened state. In this state, the battery voltage is low and current drive capability is also at it's lowest.

An alternative to the bipolar LDO is the CMOS LDO. Figure 9.4b shows a simplified diagram of a CMOS LDO. In this illustration, the core of the device is again a bandgap voltage reference, which is buffered with an operational amplifier. Although this device is completely designed in a CMOS process, a primary difference in performance between the CMOS LDO and bipolar LDO is the output pass transistor. With the CMOS LDO, the dropout voltage is smaller. As compared to the bipolar PNP pass transistor device where the dropout voltage is approximately 0.3 V, the CMOS LDO dropout voltage is dependent on the $R_{DS\text{-}ON}$ of the MOSFET pass transistor. When this type of device is driving zero current out of the OUTPUT pin, the dropout voltage is also zero volts. With increased output load current, the dropout voltage will increase. For instance, if the MOSFET $_{RDS\text{-}ON}$ resistance is 0.33 ohms and the output load current is 100 mA, the drop-out voltage is 33 mV. As an added benefit, the output transistor is capable of driving currents of any magnitude without an appreciable increase in ground current.

You might ask why are bipolar LDOs still existence? The bipolar LDO offers some features that the CMOS device has yet to match. The bipolar LDO generally has lower output noise. CMOS LDOs sometimes have a noise reduction pin, as do the bipolar LDOs. A capacitor is between ground and the noise reduction pin to reduce noise. But the CMOS LDO noise can't be reduced to the extent of their bipolar cousins. The difference in output noise of the

CMOS LDO compared the to bipolar LDO is in the range of 10x, bipolar LDO noise being the lowest. In noise sensitive, precision circuits this may be a guiding factor in the LDO selection process. Additionally, the bipolar LDO is generally more effective in circuits that have high-voltage inputs and requires large output currents. Although the high-current driving/high-voltage capability of the bipolar LDO is better than the CMOS version. This is usually not a preferred feature in battery applications.

Looking at this from a different perspective, CMOS processed LDOs have the added benefit of possibly having more digital functionality. For instance, it is not unusual for this type of LDO to have shutdown capability. This feature comes in handy when you have a dual CMOS LDO in the package. Additionally, some CMOS LDOs have a low-power supervisor option where the user is told when the voltage level of the battery drops below a safe region. This supervisor option implements the full functionality of the standard supervisor stand-alone chip, including a time-out timer function on a RESET pin along with transient rejection. This feature is very effective when reduction of chip count is required in the small, compact battery powered application.

Comparing the Three Power Devices

Efficiency of these types of power circuits can be defined as the ratio between output power and source power, or

$$\% \text{ Efficiency} = 100 \times ((V_{OUT} \times I_{OUT}) / (V_{SOURCE} \times I_{SOURCE}))$$

With this formula, a snapshot shows the effectiveness of the power supply under specified conditions. A study of this type of snapshot is interesting; however, the changes of efficiency as it relates to changes in the source voltage and output current give the guidance that is needed for insuring that the circuit design is optimized.

Figures 9.5 and 9.6 illustrate the results of efficiency experiments with these three types of devices. In both of these figures, data was taken using the TC105 Buck-SPC, TC1185 LDO, and TC7662A charge pump (from Microchip). The TC105 is a step-down Buck-SPC that provides output currents up to 1A(max). This device normally operates in a PWM mode, but automatically switches to a pulse frequency modulation (PFM) mode at low output loads for greater efficiency. The TC7662A charge pump voltage inverter converts voltages spanning from 3 V to 18 V to –3 V and –18 V. This device has an onboard oscillator and can source output currents as high as 40 mA. The TC1185 is a CMOS LDO. Using a CMOS LDO as opposed to a bipolar LDO eliminates wasted ground current, thereby increasing the efficiency. This device is stable with an output capacitor of 1 µF and capable of 150 mA (max) output current.

Figure 9.5 shows a plot of the efficiency of these three devices, with respect to the source voltage. As shown, the TC105 Buck-SPC is best in class with the TC7662A charge pump a close second. It is important to note that the charge pump is unregulated so its efficiency curves can be misleading. If you evaluate a regulated charge pump with these same conditions, the

Chapter 9

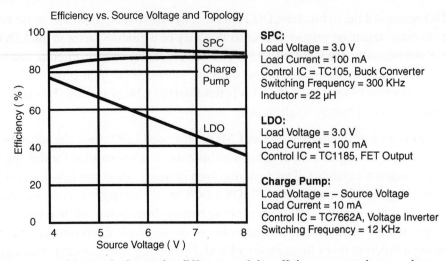

Figure 9.5: This graph shows the difference of the efficiency versus input voltage of SPC, LDO, and charge pump circuits. It illustrates that an SPC device is the most efficient device, when the source voltage is dynamic. The LDO device is the least efficient circuit in a system, showing a decreasing linear relationship of efficiency with respect to the source voltage.

Efficiency versus Source Voltage curve would degrade with increased voltage. In contrast, the TC1185's (LDO) efficiency degrades with increased input voltage linearly, as is expected.

Figure 9.6 has an evaluation of these same devices. In this figure, the results of their efficiency are looked at with respect to output current. Figure 9.6 shows that the SPC is very efficient

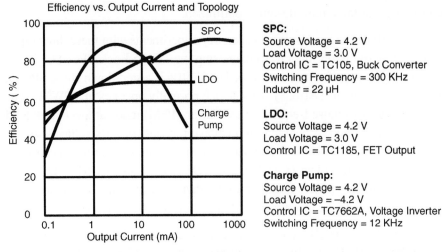

Figure 9.6: The graph illustrates the difference between the efficiency versus output current of SPC, LDO and charge pump circuits. This data demonstrates that the conditions of the specific system dictate the optimum solution.

over a wide range of load current as well as a wide range of input voltage. For applications that have a wide input voltage and over 100 mA load-current requirement, the SPC is a clear winner in efficiency.

What is the Best Solution for Battery-Operated Systems?

A typical challenge for the power-management scenario, that uses the types of devices referenced above, is the battery-powered application. In particular, an application that uses one Li-Ion, rechargeable battery cell and requiring an output voltage of 1.8 V and output current of 0 to 300 mA can be found in a large variety of battery applications. The nominal voltage of a Li-Ion cell is 3.6 V with an output voltage range from 2.8 V to 4.2 V. This is one example of a battery-powered application. For this application, you can evaluate the feasibility of the SPC, charge pump and LDO solutions in terms of efficiency.

The charge pump is probably the most unlikely device to fit in this application. The primary reason that TC7662A is not a good fit is because it is an inverting device. In other battery applications, you may need this negative power supply. But, for this application, the TC7662A would not be able to easily provide a 1.8 V output. Although this is obviously a very critical characteristic, the charge pump has other problems in this application. As shown in Figure 9.6, the Efficiency versus Output Current performance of this device is only optimum for output currents from 1 mA to 10 mA, which would infer that it would be better suited for lower output current applications. Although the charge pump is inexpensive, it can exhibit high $\delta I/\delta t$ conducting emissions. Additionally stray-lead, trace, and package inductance can become sources of shielded radiated emissions. The efficiency of this type of device does not suit this application well.

The LDO is a possible solution to this application problem. The implementation of the TC1185 in a circuit is very simple due the low, external device-count. Although the TC1185 can be designed into this application, the Efficiency versus Source Voltage is not as good as the other solutions. Since the battery has a large range of output voltage, the frequency of battery charging is higher with the LDO as opposed to the SPC solution. In fact, the LDO power dissipation is a total of 240 mW when the input voltage is 4.2 V and the output is loaded to 100 mA. From a positive perspective, the TC1185 does not emit any EMI signals, which may be desirable in your application.

The SPC device is the best choice for this type of application problem, if EMI is not an issue. The TC105 can easily power a 1.8 V output with good regulation. The Efficiency versus Source Voltage performance of this device is comparatively very high with approximately a value of 90% efficiency over the entire source voltage-range as shown in Figure 9.5. In addition to this Efficiency versus Source Voltage, the Efficiency versus Output Current is not superior at every current as shown in Figure 9.6, but it remains the best choice across the entire output current range.

Chapter 9

In battery-powered applications, efficiency considerations are at the top of the list when the power-management strategy of new design is considered. The choices for this application problem are SPC, charge pump and LDO. For systems that have wide variations over the source voltage, the SPC is by far the best choice for this application. If the system operates over a small range of low-output currents, the charge pump can provide the best efficiency in a system. And finally, if the system requires a good, low noise regulated output and the power dissipation is manageable, the LDO will provide satisfactory results.

Designing Low-Power Microcontroller Systems is a State of Mind

Low power is a "state of mind." Think about it. You can throttle down your controller to near inactivity if you really want to save battery power. The straightforward way to turn the heat down is to re-evaluate clock sources or reduce the power supply voltages. Another equally effective approach is to operate with a partial or complete controller/processor shutdown mode. But, if you combine these techniques with execution time and a little intelligence, you can easily tackle your most challenging power conservation problems.

Making Analog and Digital Play Together

You can accomplish an improvement in power consumption and consequently an increase in functionality if you understand the tools available in the microcontroller as well as hardware options. One dimension of power conservation is controlling the magnitude of the power-supply voltage in your application. I would imagine that you are interfacing to the real world at some point during your code operation. If you are, you will have analog content in your circuit. Analog power supply requirements are higher than the requirements for digital. In addition, don't forget that analog noise margins are much smaller than digital noise margins. The analog noise floor does not reduce with lower, power-supply voltages. It stays the same over power-supply-voltage changes. For example, a 12-bit ADC can produce good, solid conversions with a 5 V supply. However, that same 12-bit ADC will produce a smaller number of noise free bits when you use a 2 V supply. This is because the LSB size has become smaller, but the magnitude of the noise is consistent. The solution to this problem is to use higher supply voltages when running analog and lower voltages during the digital-only operation.

The diagram in Figure 9.7 shows a simple, microcontroller, battery-operated system. This circuit uses the PIC18F1320 from Microchip. The PIC18F1320 has features, such as a variety of idle modes and a two-clock start-up capability, which can enhance your low power strategy.

On the hardware side, industry is continuing to develop classes of external peripherals as well as the internal microcontroller peripherals with lower power performance in mind. In terms of the external peripherals to the microcontroller, you can achieve lower power by reducing power-supply voltage requirements to the chips and optimizing topologies for the lower-power jobs. This simple example (Figure 9.7) has low-power operational amplifiers, A/D converter, and a regulated-adjustable charge-pump.

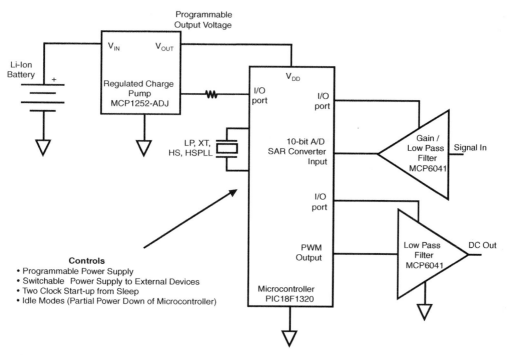

Figure 9.7: The lower-power, external-peripherals to the microcontroller only provide half of this battery-powered system implementation. The programmable capabilities of the microcontroller allow control of the power-supply voltage in the system (MCP1252-ADJ) and control of the microcontroller system clock and sleep modes.

With external or internal integrated ADCs, the amount of power dissipated is more dependent on the converter topology than on IC design innovation. For example, the ratio of conversion time to current consumption of the successive approximation register (SAR) converter is considerably lower than the sigma-delta converter. You will probably use the SAR converter in battery-powered applications, unless you need higher resolution and accuracy.

The power supply in the circuit in Figure 9.7 is adjustable. A higher voltage of 5 V is best suited for analog circuitry and a lower voltage of 2 V is best suited for digital activities. The adjustable power-converter in Figure 9.7 has high efficiency with low-output currents and Li-ion battery input voltages (4.2 V down to 2.8 V). For these reasons, this circuit uses a regulated, adjustable charge pump, DC/DC converter (MCP1252-ADJ).

Controlling the power-supply voltage for various operations is only half of the story. If you really have a lower power "state of mind," you will want to power down some parts of the controller while letting other sections continue to operate. As an example, you can independently run an A/D or D/A conversion or the USART communication interface from the controller. These chip functions may only need power locally.

Chapter 9

Optimizing the external peripheral-power trade-offs is also important. You will also find real power savings when using the external and internal peripherals in concert with the microcontroller programming capability. For instance, the microcontroller controls the power-supply voltage by switching a new configuration into the resistive feedback system of the MCP1252-ADJ regulated, adjustable-output charge-pump. The charge-pump generates a higher output voltage to insure that the analog circuitry performs at its optimum level. Digital events from the microcontroller can tolerate a lower power-supply voltage. For instance, the power-supply specifications of the PIC18F1320 are from 2 V to 5.5 V. You can calculate the power savings for this type of change as a direct ratio of the two voltages from the charge pump. As an added benefit, if the external peripherals are powered down with the lower-power supply voltage using the I/O ports. In this manner, the power saving are further improved.

Controlling Your Clocks

One issue that is often overlooked when designers are trying to reduce the overall power consumption of an embedded system circuit is the management of the clock when the controller comes out of its sleep mode.

A microcontroller or microprocessor can have a variety of clock sources (Figure 9.8). The most obvious clock-source is an external one. In this instance, you would connect a crystal, resonator, an internal controller clock or a clock generator to the appropriate device pin. Beyond

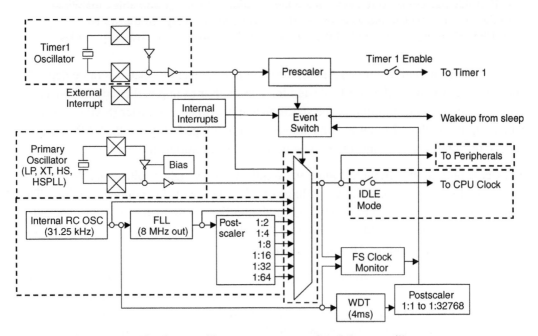

Figure 9.8: This diagram illustrates an example of three oscillator sources for a microcontroller. Externally, there are two clock connections (Timer1 and primary oscillator). Internally, there is another oscillator.

these elements, that generate the clock signal, the microprocessor or microcontroller can have a postscaler, prescaler, or frequency lock-loop (FLL). The postscaler and prescaler divide the input clock frequency down. You would use the FLL to multiply the input clock frequency.

In a real time operating system, it is critical to have a clock management strategy when the system wakes up for short periods of time and sleep for a long period after the wake-up. If the wake-up time is typically <1 sec and you are using a crystal oscillator or ceramic resonator, you may find that there will be a delay between pulling out of the sleep mode and beginning to execute code. The microcontroller or microprocessor will not execute code during this delay or start-up time. However, the application circuit will be consuming power.

For example, Figure 9.9 shows the typical start-up time for a 4 MHz crystal oscillator. In Figure 9.9, this time is approximately 450 msec. If this crystal oscillator were the only clock connected to the controller with a one second code-execution time, the actual execution time of the code would be 45% longer than expected. During the clock start-up time, your circuit is consuming power but not executing code.

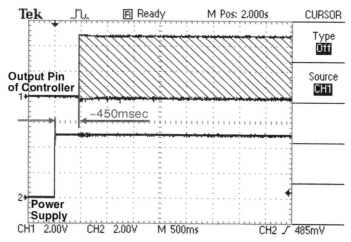

Figure 9.9: The clock connected to the controller is a 4 MHz crystal. In this oscilloscope plot, Channel 1 is a microcontroller output pin. The programming of the microcontroller toggles this output- pin every 12 instruction cycles. Channel 2 shows the power-supply to the controller, V_{DD}. The external clock to the controller becomes active approximately 450 msecs after power supply start-up.

With this type of application, it might be advisable to use an internal clock to execute code. The internal clock will start up nearly instantaneously. It is not unusual a 4 MHz internal clock to start in a few microseconds. Figure 9.10 illustrates the start-up time of an internal clock.

This is ~50,000× improvement from the 4 MHz crystal oscillator. From this data, one might summarize the appropriate clock for this type of application is an internal clock. The power consumption of internal clock is nearly equivalent to the power consumption of a crystal oscillator. This strategy works as long as your controller or processor is not required to run time-critical operations, such as USART communications or timing a precision pulse.

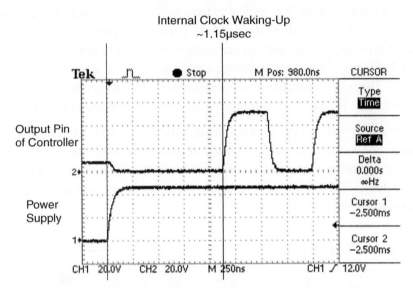

Figure 9.10: The start-up time of the internal clock is approximately 1.15 µsec. This is considerably faster than the start-up time of the oscillator (in Figure 9.9).

A third clock source that you may be evaluating is the resonator. Figure 9.11 illustrates the start-up time of a resonator.

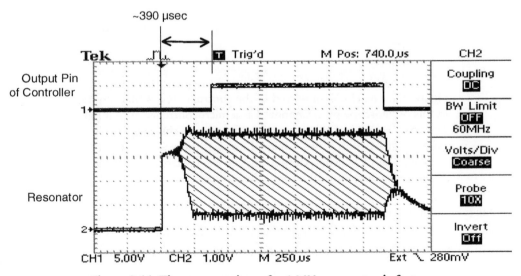

Figure 9.11: The start-up time of a 4 MHz resonator is faster than a 4 MHz crystal and slower than a 4 MHz internal clock.

Systems Where Analog and Digital Work Together

There is a clocking system that you can use that is better than any of these three clocking options. The best of all worlds is to quickly determine if your circuit needs a precision clock. If the controller needs a precision clock, the oscillator or resonator is turned-on. If not, the controller will shut down. You can make this determination quickly, after the controller leaves its sleep mode. If you combine the internal clock with an external resonator or crystal oscillator, you can quickly make this type of decision. You can find significant improvements in power consumption when you use two clock sources (instead on one).

This technique is called the two-clock start-up strategy. In this hardware/firmware configuration, the controller uses two clocks. Both clocks are off during the sleep mode of the application. At the time of wake-up, the internal clock is turned-on to quickly determine if the crystal oscillator is required. If the crystal oscillator is required, the internal clock continues to execute code until the crystal oscillator is up and running. At this time the controller switches over to the crystal oscillator and turns off the internal clock.

Working the Digital Angle with Sleep Modes

The central focus of a successful low power design is a microcontroller that has a variety of sleep modes and clock modes. You can conserve system power with the idle modes and sleep modes of the microcontroller. The idle modes of the microcontroller power down the CPU while allowing functions such as the 10-bit A/D converter to continue to operate. The sleep mode implements a complete shutdown of the controller.

When the clock of your controller switches states, the various, controller, logic-gates in your controller pulls current from the power source. When looking at your current consumption in your controller, the first stop is to look at the clock power consumption. If you examine the types of clocks you can use, the internal clock will run with less power than the frequency equivalent crystals, oscillators, or resonators.

Some controllers/processors have three fundamental modes of operation. The first is the full bore run mode where every thing is up and running. An intermediate mode is the idle or wait mode where the peripherals are usually running but not the controller. The third, and most important mode for lower-power, battery operation is the sleep or stop mode. In this mode, the device stops consuming power completely. The sleep mode generally disables the system's clocks but power conservation is more effective if you also disable the external clock sources.

Here are some additional suggestions to complete your low power strategy. Drive any unused I/O pins into a high or low state. Use internal clock where possible. They generally are the lower power choice. Shut down all peripherals not in use, like the pulse width modulator (PWM), A/D converter, USART, etc. Use as many lookup tables as possible in your code instead of using the CPU to compute results. Check the power consumption of all external components. For instance, measure the voltage drop of *all* external resistors in the circuit. Lower the I/O pins that are used to power external peripherals such as serial EEPROMs or external analog devices. Another surprise can be your LEDs that are turned-on. A single LED can wipe out your power savings efforts. In general, look for current consumption gremlins.

Conclusion

Device power-savings in battery-powered applications is extremely important. You will achieve true value by using the microcontroller's programmability. You can do this by changing the power-supply voltage at the output of a regulated charge pump. A second area would be to power-down noncritical peripherals when not in use. Another option is to control the clocking strategy in order to optimize power versus functionality. Integrated circuit manufacturers are continuing to improve the dynamic performance of their peripheral devices while reducing the quiescent current and supply voltage requirements. Microcontroller manufacturers are adding modes, such idle modes and sleep modes, that save average power over long periods of time. The combination of lower-power peripherals and microcontroller modes enhance the chances of having a low-power, battery-powered solution.

Got your checklist? Now take all of these variables and put on your low power "state of mind" hat. You, as the perceptive programmer/hardware expert, will have to evaluate each one of your applications, every situation inside those applications and look for the power consumption gremlins. Good luck!

Chapter 9 References

Portable Electronics Product Design and Development, Haskell, Bert, McGraw Hill, 2004.

The Firmware Handbook, Ganssle, Jack, Newnes, 2004.

Power Supply Cookbook, Brown, Marty, Newnes, 2001.

CHAPTER 10

Noise – The Three Categories: Device, Conducted and Emitted

CHAPTER 10

Noise – The Three Categories: Device, Conducted and Emitted

Are your circuits unstable? Or do they tend to give you different results from one moment to the next? Too much noise is a typical problem confronting many circuit designers. In this chapter, noise is defined as undesirable signals that are present in a circuit. This definition excludes analog nonlinearities, which may produce distortion. Once you evaluate where your noise sources are, eliminating circuit noise can be quite simple. Tools such as filters or lower noise devices provide effective solutions. Your circuit board layout will also be critical. If you want to learn about layout techniques, refer to Chapter 11.

Figure 10.1 shows three primary types of noise found in analog applications. Each type of noise has its own set of possible solutions. The first noise type is device noise. Device noise is the intrinsic noise of the devices in the circuit. Examples of device noise would be the thermal noise of a resistor or the shot noise of a transistor. Another type of device noise is the switching noise from a switched mode power supply (inductive based) or a switched capacitor converter (capacitive based).

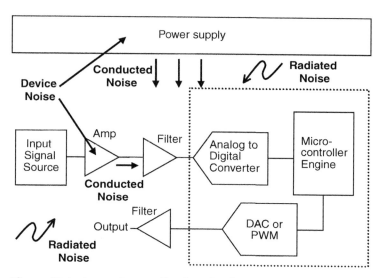

Figure 10.1: An analog application circuit can acquire noise from a variety of sources. You can categorize these noise sources into three groups: device noise, radiated noise, and conducted noise.

233

A second type of noise is radiated noise. You will find radiated noise originates with EMI sources, such as switches or motors. These sources couple radiated noise into the signal. Radiated noise can also transmit from trace to trace on your PCB.

The third type of noise that affects the performance of analog devices is conductive noise. This type of noise already exists in the conductive paths of the circuit, such as the power lines or signal path. Conducted noise mixes with the desired electrical signal. The origin of conducted noise is either device noise or radiated noise. In this chapter we will primarily cover device and conducted noise, leaving the discussion of radiated noise for Chapter 11, "Layout/Grounding (Precision, High Speed, Digital)."

Definitions of Noise Specifications and Terms

Noise in electronics can be random or connected to some circuit generated frequency. If it is a random event over the frequency spectrum, it is void of coherent frequencies. Based on your knowledge of the input of your circuit, you cannot predict these types of noisy events. They occur inside all analog devices, including passive and active devices. If you sample these noise events, they will build a normal distribution over time as illustrated in Figure 10.2. If sampled noise events do not have a normal distribution, the prediction of peak-to-peak over time is difficult to do. If the noise falls into a normal distribution, you can apply mathematics in order to bound and describe the apparent noise randomness. We will discuss the type of circuit noise that does not fall into a normal distribution later in this chapter in the "Power Supply Noise" section.

If the noise samples fall within a normal distribution, repeated samples differ around a central value. The distribution is roughly symmetric around this central value. The distribution produces a curve with its highest occurrence at the center point, tailing off to zero in both directions. Because this distribution is consistent with the Central Limit Theorem, you can use standard calculations such as mean and standard deviation, to predict the general magnitude of future occurrences with respect to the normal curve.

There are three diagrams in Figure 10.2. Each shows a different representation of the same data. The first diagram, Figure 10.2a, is a time-based oscilloscope picture of a noisy signal. This signal is void of coherent frequency signals and it seems to be random. This picture of noise may be similar to what you have seen on the bench, particularly if you zoom in using your magnitude/div and time/div knobs.

The noise in Figure 10.2a is sampled and mapped into the magnitude-based graph in Figure 10.2b. In this graph, the number of samples is on the y-axis and the sample magnitude on the x-axis. The time scale from the previous diagram is lost and only the magnitude of each sample is preserved. It is interesting that the samples from Figure 10.2a build a normal distribution in the histogram plot (Figure 10.2b). This is not unusual because of the random nature of noise. With this sampled data, you can calculate the root-mean-square (rms)

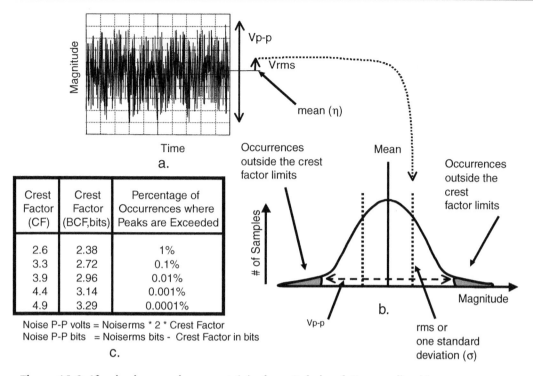

Figure 10.2: If noise is a random event (a), these "obviously" unpredictable events can be characterized with a normal distribution (b) where you can calculate mathematical equivalents of the mean (η) and standard deviation (σ). If you multiply the standard deviation (equal to rms by a 2x crest factor, you can predict the probability of occurrences outside the crest factor limits (b and c).

value (standard deviation) and mean of the sample set. And if this histogram forms a normal distribution, you can exercise a multiple (called a *crest factor*, Figure 10.2c) to the rms value to establish peak-to-peak noise values for your data. This calculated peak-to-peak value will predict the percentage of future samples that will fall outside those limits.

It is important to know that when you use the noise calculations they should be treated as best estimates, rather than absolutes. This may give you a feeling that you can't count on any of the results that you arrive at, but to the contrary. This is where statistics come into the picture. These statistical estimates provide a degree of confidence that you will consistently get the expected results, if in fact, your sample is a good representation of the population.

Evaluating Noise with a Circuit Example

Take for example the circuit in Figure 10.3, and the test results in Figure 10.4.

In Figure 10.3, a load-cell circuit accurately measures the weight applied to the sensor. With a 5 V excitation voltage applied to the high side of the sensor, the full-scale output swing is

Chapter 10

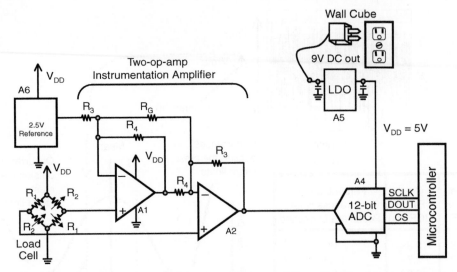

Figure 10.3: The two-op amp instrumentation amplifier gains the DC signal at the output of the load-cell sensor. The 12-bit A/D converter then digitizes that signal.

a +/-10 mV differential-signal with a 32 ounce maximum excitation. This small differential signal is gained by a two-op amp instrumentation amplifier (G = 153 V/V). I chose a 12-bit converter to match the required precision of this circuit. Once the converter digitizes the voltage presented at its input, the microcontroller receives the digital code through the SPI™ port. The microcontroller then uses a look-up table to convert the digital signal from the ADC into load-cell weight.

Figure 10.4 shows how the circuit noise can contaminate a perfectly good A/D converter. In this figure, the 12-bit converter was used to unsuccessfully convert a DC signal. Initially, one might assume that the 12-bit converter is not very good. But it is possible, as is with the

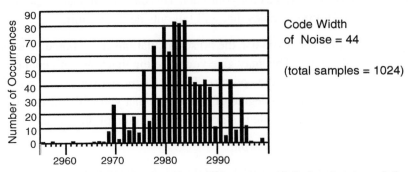

Figure 10.4: A 12-bit A/D converter is used to convert DC signal to hopefully one code. If you don't manage the passive (from resistors) and active (from amplifiers) noise well, the converter will give an unrepeatable result over time. The LSB size in this graph is 1.22 mV. The code width of forty-four codes equals 53.68 mV.

circuit, that the active and passive devices in the signal path are noisier than expected. The components I chose for this first circuit were:

$R_3 = 300\ k\Omega$, $R_4 = 100\ k\Omega$, $R_G = 4020\ \Omega$, (+/–1%)
Single-supply op amps with input voltage noise density = 29 nV $\sqrt{}$ Hz @ 1 kHz

Figure 10.4 shows the results of ignoring noise issues in this circuit. The sampling rate was 10 ksps for this data. There were 1024 samples taken for the histogram. I took care to insure that the input signal was DC or relatively noiseless. The noise spread of my data is 44 codes wide. I was surprised to find that this code width was repeatable (± 2 codes)! Since this data was taken using a 12-bit A/D converter, which has 4096 possible output code combinations with a 5 V reference, one LSB equals about 1.22 mV. If you translate the code-width of 44 into millivolts, this would equal approximately 53.68 mV of noise. You will note that the data in Figure 10.4 doesn't really have a normal distribution. Consequently, I suspected that there is a frequency component embedded in my data.

Then I went back in an effort to reduce the noise so that my converter would provide a single conversion value, every time. Figure 10.5 shows this modified circuit. It is not a coincidence that you find the circuit in Figure 10.5 in Chapter 3 (Figure 3.8). This is a common circuit solution for pressure sensing applications for good reason. One of the reasons is that it produces an accurate, low-noise solution.

In this modified circuit I reduced the resistor values (without changing the function of the circuit) and replaced the operational amplifiers with a lower-noise version. These modifications

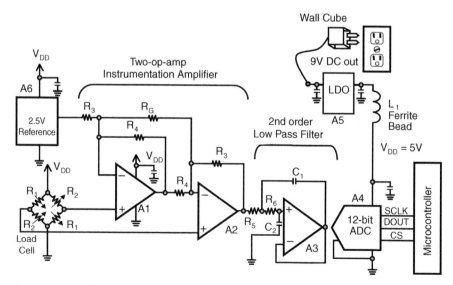

Figure 10.5: This modified circuit is basically the same as the circuit in Figure 10.3. The modifications include the addition of a low-pass, anti-aliasing filter, bypass capacitors and an inductive ferrite bead (L_1) on the power supply trace.

Chapter 10

attacked my device noise problems. From there, I added a ground plane, which is highly recommended in any analog circuit. This was done to address emitted noise problems and as well as conducted noise problems. I proved that emitted noise was a problem by moving the board around my lab. Sometimes I would place it closer to the wall cube and other times closer to my fluorescent lights. This would increase the circuit noise in both places. I also found that when the copy machine was working in the cubical next to me, I had an increase of noise.

I finally tackled my conductive noise problems by inserting an analog low-pass filter after the instrumentation amplifier. I also put a "choke" inductor (L_1) on the power supply trace and installed appropriate bypass capacitors.

The components I chose for this second pass circuit were:

$R_3 = 30$ kΩ, $R_4 = 10$ kΩ, $R_G = 402$ Ω, (+/–1%)
Low-pass filter – $R_5 = 27.4$ kΩ, $R_6 = 196$ kΩ, $C_1 = 100$ nF, $C_2 = 470$ nF
Single-supply op amp with input voltage noise density = 8.7 nV $\sqrt{}$ Hz @ 10 kHz
Bypass capacitors = 0.1 µF on every active device

This type of attention-to-detail paid off. As Figure 10.5 illustrates, the converter is perfectly capable of converting to one bit in a reliable and repeatable manner.

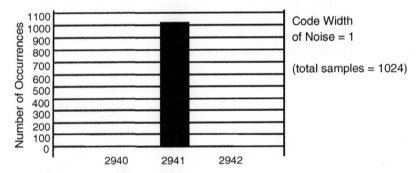

Figure 10.6: This data uses the same 12-bit A/D converter to convert DC signal to one code. In this new circuit, I am reducing the resistor values and using lower noise operational amplifiers. These two actions reduce the device noise in the circuit. An added ground plane reduces emitted noise. And finally, I add a low-pass filter, bypass capacitors, and a power supply inductive filter to reduce the conducted noise.

These results are impressive, but you may be curious about how I know that these changes were the right ones. So let's dig into each type of noise is some detail.

Device Noise

Devices come in many flavors, each with their own frequency spectrum of noise. The passive and active devices, all capable of creating noise, can have an impact on your circuit. The passive devices include resistors, capacitors, and inductors. Of these three types of passive devices, the resistor is most notable for circuit noise. This type of noise can be a nuisance. Capacitors generate random noise that is dependent on temperature and Boltzsman's constant, equally KT/C (where K is Boltsman's constant, T is temperature in Celsius, C is capacitance in farads). Capacitors and inductors can also reduce noise in a signal path or on the power supply lines. The capacitor and inductor also create noise if there is some kind of switching signal going across them. So we will discuss these passive devices as they relate to switch-mode power supplies and charge pumps.

Another source of device noise comes from active devices. The typical active devices that will be in your circuits are operational amplifiers, A/D converters, voltage references and power supply chips. Of these devices, the operational amplifier has the largest direct effect on the signal chain and the power supply chips will inject noise through the device's power supply pins, which may eventually enter the signal chain.

Resistor Noise

An ideal resistor will create noise inside the resistive element. This type of noise is commonly called thermal noise or Johnson noise. The resistive element generates noise because of the agitation of particles in the resistor. You cannot avoid resistor noise by not applying power. This noise will appear across the resistor whether it is powered with a current source, voltage source, or not at all. This noise can however be filtered and consequently reduced later on down the signal path as conducted noise. The ideal resistor will create a predictable noise that is flat across the full frequency spectrum. The ideal resistor rms noise is equal to:

$R_{NOISE} = \sqrt{(4 * k * T * R * BW)}$

Where k is Boltzman's constant equal to $1.38*10^{-23}$

T is temperature in Kelvin
R is the resistor in ohms
BW is the bandwidth of interest

In this calculation, Kelvin is also equal to 298.16°K (25°C) at room temperature. Each degree increase in temperature in Kelvin is equivalent to a 1 degree increase in Celsius.

With this calculation, it is easy to quickly determine if your resistor is too noisy for your circuit. For instance, the noise of a 1 kΩ resistor at 25°C is approximately 4 nV / √Hz (rms). If you want to calculate the amount of noise generated by an ideal 1 kΩ resistor across a frequency bandwidth of 1 Hz to 1000 Hz, that noise would be equal to:

$V_{NR} = \sqrt{(4 * k * T * R * BW)}$
$V_{NR} = \sqrt{(4 * 1.38 * 10^{-23} * 298.16 °K * (1000 Hz - 1 Hz))}$
$V_{NR} \sim 126$ nV rms
$V_{NR} \sim 834$ nV p-p (assumes a crest factor of 3.3, see Figure 10.2)

Chapter 10

This doesn't seem like very much noise. For example, if your system uses a 16-bit A/D converter with a 5 V reference (5 V FSR), the LSB size with this converter is 76.3 µV. The question is: "If I have an input resistance of 1 kΩ to my converter, will it affect my signal accuracy?" The answer is no. Our peak-to-peak estimate is equal to 1% of one LSB. On the other hand, if you are designing with a 20-bit A/D converter (5 V FSR), our peak-to-peak estimate is equal to 17.5% of one LSB. Now, that very well may be a problem!

Table 10.1 gives the ideal resistor noise for a variety of resistors.

Table 10.1: All resistors generate noise. This table tabulates the ideal room temperature noise that a variety of resistors will generate.

Table of Resistance Noise @ 25°C (298.16°Kelvin)

Resistance (Ω)	Noise Density (nV/rt Hz)	Resistance (Ω)	Noise Density (nV/rt Hz)
1	0.1283	100	1.283
2	0.1814	200	1.814
3	0.2222	300	2.222
4	0.2566	400	2.566
5	0.2869	500	2.869
6	0.3142	600	3.142
7	0.3394	700	3.394
8	0.3629	800	3.629
9	0.3849	900	3.849
10	0.4069	1000	4.069
20	0.5737	2000	5.737
30	0.7027	3000	7.027
40	0.8114	4000	8.114
50	0.9072	5000	9.072
60	0.9937	6000	9.937
70	1.0734	7000	10.73
80	1.1475	8000	11.48
90	1.4813	9000	14.81
100	1.2829	10000	12.83

The units of these ideal calculations are nV/√Hz (rms). The values in this table are easy to convert to higher of lower resistors that are not on the chart. This is done by multiplying or dividing the quantity by the square root of the ratio between the table value and the value of interest. For instance, if you want to know what type of noise a 1,000,000 Ω resistor will create you need to multiply the noise of the 10,000 Ω resistor by 10, or √(1,000,000/10,000). If you work with these ideal calculations, you will generally be able to determine if the resistors in your circuit are causing noise problems.

Real resistors, like wire wound, film type or composition resistors produce noise that is higher than the ideal. Of these three types, the wire wound resistor is the quietest followed by the film type with the worst being composition. A good quality wire wound can produce noise that is nearly ideal. On the other end of the spectrum, the composition resistor will generate noise because of its contacts. This noise is a result of the individual particles in the contact. Contact noise is proportional to the DC current flowing through the resistor. This noise will appear at lower frequencies and looks very similar to the 1/f noise of amplifiers. A composition resistor that has no DC current conducting through it will exhibit near ideal noise behavior. Film type resistors also have contact noise, but to a lesser degree. This is primarily because the film resistor contact is made of a more homogenous material. Variable resistors, such as analog and digital potentiometers exhibit the same types of noise as described above and don't forget the additional wiper resistor noise. With all of these types of resistors, they will operate at a lower noise level if you keep them under their power rating.

Figure 10.7 shows the ac model for a real, wire-wound resistor. The parasitic capacitor (C_p) will have the most impact on the noise behavior (as well as the frequency response). This is because it attenuates higher frequency noise. This is a nice benefit, but you should still use caution when using higher value resistors. They will still have higher, low-frequency noise.

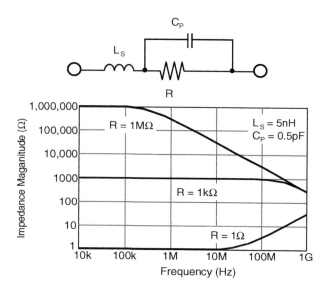

Figure 10.7: The non-ideal resistor can be modeled using an ideal resistor, R, a parasitic capacitor, C_p, and series lead inductance, L_s. These parasitic elements attenuate the resistor noise at higher frequencies, particularly with higher value resistors.

Revisit Our Application Circuit With Better Resistors

Going back to the circuit in Figure 10.3, I have run a SPICE simulation to try to determine whether or not a reduction in resistor values around the instrumentation amplifier will make a difference. I know that I need these resistors and I know that the output of these amplifiers should have at least a 1000 Ω load. It is easy enough to reduce all of these resistors by 10× or 100× without changing the gain of the instrumentation amplifier circuit.

Chapter 10

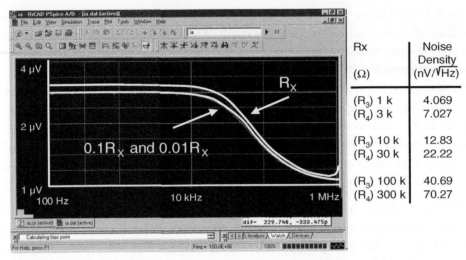

Figure 10.8: This is a SPICE simulation of the instrumentation amplifier using three different sets of resistors. The x-axis is a log scale of frequency in Hertz and the y-axis is the magnitude of noise divided by the square root of Hertz. The top curve, R_x, has $R_3 = 100$ kΩ and $R_4 = 300$ kΩ. This combination of resistors is the noisiest. The curves 0.1 R_x and 0.01 R_x are on top of each other. This is because the amplifiers in the instrumentation amplifier start to dominate the noise response. With these two curves, the circuit noise is lower than the R_x curve.

An easy way to determine whether or not I have resistor noise is to run a SPICE simulation (see Figure 10.8). In my simulation, I have made sure that the DC simulation is correct. In particular I am looking for extraordinarily unexplained high voltages, input amplifier nodes that are not where they are suppose to be and an output voltage on A_2 that is in between the supply voltages. After this reality check, I have more faith in the AC analysis of Figure 10.8. For more details about SPICE simulations, refer to Chapter 7, "SPICE of Life."

In this figure, I ran the circuit SPICE simulation with the resistor values called out in Figure 10.3. The second and third AC simulation replaced all of the resistors with 1/10 of the values and then 1/100. As you can see in this figure, resistors that are 10x lower buys me some reduction in noise. But not as low as I would expect. By lowering the resistors by 10× I should see an improvement in noise equal to $\sqrt{(1/10)}$ or a 32% reduction. This did not happen. And further more, reducing them by 100× buys me nothing. This suggests that my amplifier is the limiting factor.

Operational amplifiers

It is important to understand the noise that operational amplifiers create because almost every analog device will have an op amp somewhere in the circuit. The op amp noise behavior, over frequency, has a signature that is unmistakable.

If you look for the amplifier noise specification in the typical amplifier data sheet, you will notice that it is a "referred-to-input" specification. The location of this noise source is at the noninverting input of the amplifier. In the specification table, you will typically find input noise and input noise density specifications. The input noise specification will describe the low frequency noise of the amplifier in terms of a bandwidth. You will find this bandwidth in the "conditions" column. 1/f noise is this lower frequency noise. This is mainly because this part of the curve actually follows the ratio of 1:frequency times a multiple. The transistors in the input stage of the amplifier generate the noise through this frequency band. This is primarily the differential input stage, but it also includes the input stage load transistors.

Input noise density calls out a noise figure that refers to one frequency. For instance, the noise specifications in Figure 10.9 identify the input voltage noise density at 10 kHz to equal 8.7 nV/√Hz. You measure the input voltage noise density at the specified frequency across a 1 Hz bandwidth. Usually this specification appears in the broadband noise portion of the frequency plot (Figure 10.9). Theoretically, this broadband noise is flat. Assuming that it is flat is a good estimate of the amplifier's behavior. It is also the foundation or base line of the 1/f noise portion of the curve. The diffused resistors inside the operational amplifier primarily generate the broadband noise. These resistors can be diffused resistors or the source or drain of the transistors in the amplifier.

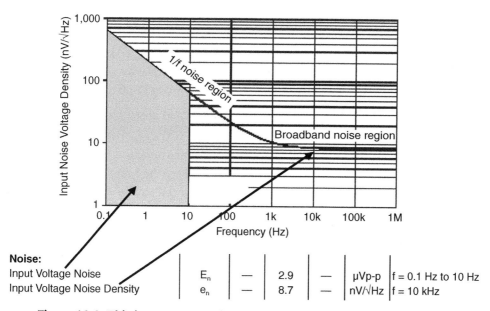

Figure 10.9: This is a representation of the noise of an example amplifier. The specifications for the noise performance of the amplifier are in tabular form at the bottom of this figure. These specifications numerically refer to the input noise voltage density versus frequency plot.

Chapter 10

Further, on in the amplifier data sheet you will find a typical specification graph that will show you the input noise voltage density vs. frequency. Figure 10.9 shows an example of this type of graph. In this example, the input voltage noise specification is equal to the area beneath the input-voltage, noise-density curve between the specified frequencies of 0.1 Hz to 10 Hz. Note that in the table the units for this specification are peak-to-peak. To convert this to an rms value, simply divide the value by 6.6 (industry standard crest factor = 3.3).

You can easily calculate the noise underneath the curve for different input voltage noise bandwidths in the 1/f region. The first order of business in this calculation is to determine the input noise density at 1 Hz. Once you find that value, the simple formula below will provide the rms noise under the curve.

$$V_{1/f : f_2-f_1} = B \sqrt{\ln(f_2/f_1)},$$
Where B is equal to the Input Noise Density at 1 Hz

As an example, the amount of rms noise produced by the amplifier shown in Figure 10.9 from 0.1 Hz to 1000 Hz is equal to:

$$V_{(1/f) : f_2-f_1} = B \sqrt{\ln(f_2/f_1)},$$
$$V_{(1/f) : f_2-f_1} = 200 nV \times \sqrt{\ln(1000/0.1)},$$
$$V_{(1/f) : f_2-f_1} = 607 \, nVrms \text{ or } 4 \, \mu Vp\text{-}p$$

When you think about noise at these low frequencies you may jump to the conclusion that you should take this formula down to a very low frequency, such as 0.0001 Hz (0.0001 Hz = 1 cycle per 2.8 hours). Be careful when you look at frequencies lower than 0.1 Hz, which is one cycle every 10 seconds. At lower frequencies, it is very possible that other things are changing in your circuit, such as temperature, aging, or component life. If you think of this realistically, low frequency noise from your amplifier will probably not appear at this sample speed. But changes in your circuit, such as temperature or power supply voltage may.

The amplifier table of specifications also gives the input noise density value. This specification is always at a higher frequency, in the area where the input voltage noise is relatively constant. For this region of the curve, multiplying the square-root of the bandwidth and the noise density derives the noise across a bandwidth. For example, if the noise of the amplifier is 8.7 nV/√Hz @ 10 kHz, the noise from the amplifier across the bandwidth of 1 kHz to 100 kHz is equal to:

$$V_{100k-1k} = (\text{Noise Density @ 10 kHz}) \times \sqrt{BW}$$
$$V_{100k-1k} = (8.7 \, nV/\sqrt{Hz}) \times \sqrt{(100,000 - 1,000)}$$
$$V_{100k-1k} = 2.74 \, \mu V \, rms \text{ or } 18.1 \, \mu Vp\text{-}p$$
Where BW is equal to the bandwidth of interest

So the challenge from the manufacture is to give you good data so you can work through the impact of their device in your application. So how do you get from the manufacture's graph to a meaningful result in your application circuit? You calculate the area beneath the noise curve and multiply that times the noise gain of the amplifier.

Let's go through this process with a real circuit and real component values.

The amplifier in Figure 10.10 is in a typical inverting gain stage. The input to the circuit is V_{IN} and the output is V_{OUT}. The voltage at V_{SS} is equal to 0 volts or ground, and the voltage at V_{DD} is equal to 5 volts. There is a 2.5 V reference connected to the noninverting input of the amplifier at the V_{REF}.

Figure 10.10 shows the internal capacitors of the amplifier. They will come into play when we start to calculate the gain of the circuit over frequency and look at the noise. C_{CM} is equivalent to the common-mode capacitance of the input stage of the amplifier. For our example, C_{CM} is equal to 6 pF. This capacitance is referenced to ground. C_{DIFF} is equivalent to the differential input capacitance of the amplifier and you will notice that it appears between the two input terminals. For our calculations we are going to use $C_{DIFF} = 3$ pF.

The parasitic capacitance of the external resistors, R_1 (C_{P-R1}) and R_2 (C_{P-R2}), are also shown in this diagram. Although you may think that these are insignificant capacitances (at ~0.5 pF), they are worth paying attention to. They may affect the noise gain of the amplifier circuit at higher frequencies. Figure 10.7 shows the frequency effects of this parasitic capacitance.

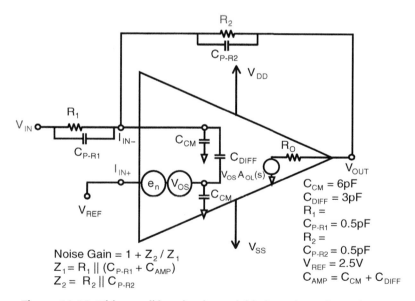

Figure 10.10: This amplifier circuit model is in an inverting gain configuration. This diagram illustrates the pertinent parasitics of the amplifier and resistors along with the calculation for noise gain.

Chapter 10

The noise gain calculation of this amplifier circuit uses the noise source, e_n, as an input signal. This source is graphically inside the amplifier symbol. You will notice that this formula is not the same as the formula for the signal gain.

Signal Gain :: $V_{OUT}/V_{IN} = -Z_2/Z_1$
Noise Gain :: $V_{OUT}/V_{IN} = 1 + Z_2/Z_1$
  Where Z_1 is the equivalent input resistor, capacitor network, and
  Z_2 is the equivalent feedback resistor, capacitor network.

When you are calculating the amount of noise that an amplifier produces, the noise gain equation will provide the correct results. This equation will also provide the correct closed-loop bandwidth of the amplifier circuit.

Figure 10.11 shows the frequency response of this amplifier circuit. The capacitors and resistors surrounding the amplifier, as well as the frequency response of the amplifier affect the bandwidth of the circuit.

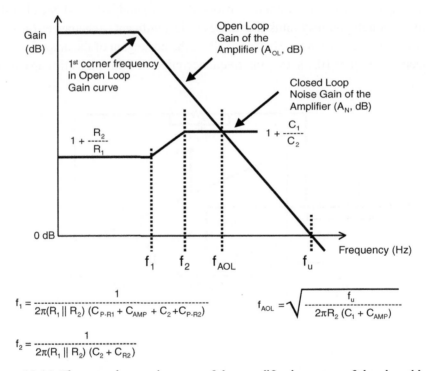

Figure 10.11: The open-loop gain curve of the amplifier is on top of the closed loop gain curve of an amplifier circuit. With the open-loop amplifier gain curve (A_{OL}), the 1st corner frequency follows the DC gain. Past this first pole, the gain of the amplifier attenuates at a rate of –20 dB/decade. With the closed-loop noise gain curve (A_N), the poles and zeros of the transfer function are shown along with their corner frequencies. The bandwidth of the noise gain is equal to f_{AOL}.

The DC noise gain of this circuit is dependent on the resistors in the circuit. At higher frequencies, the noise gain is dependent on the capacitors. It is possible in many circuits to design the second corner frequency, f_2, higher than the f_{AOL} crossing. If this is the case, you can ignore the effects of f_2. If the value of R_2 is high (>100 kΩ), f_2 may come down in frequency, lower than the open-loop gain crossing. To optimize the noise and bandwidth performance of this type of amplifier circuit, the pole, f_2, should occur at or slightly before the point where noise gain plot intersects the open-loop gain curve of the amplifier. This may require an additional capacitor in parallel with R_2.

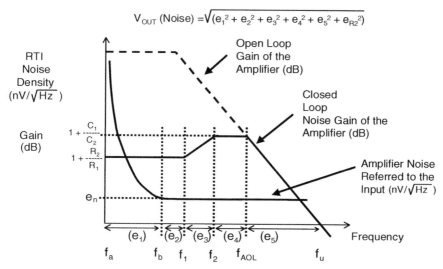

Figure 10.12: The noise of the amplifier is referred to the input of the device because the manufacturer doesn't know which configuration you are going to put your op amp into. With the referred-to-input values you can calculate your own referred-to-output values, which you can use moving forward in the rest of your circuit.

Once you calculate the gain across the frequency spectrum of the amplifier's bandwidth in this circuit, you can start to determine the circuit's referred-to-output noise. Figure 10.12 separates the noise into six parts. Five of these noise parts are in the graph and the sixth is part of the formula in the figure.

In region e_1, the 1/f noise of the amplifier is gained by the DC gain of the amplifier circuit. The specifications for amplifier noise are in nano volts per root hertz. So the analysis is complete when you multiply the average noise over the region by the square root of the bandwidth of that region. For CMOS amplifiers, the 1/f region is usually from 0.1 Hz to 100 Hz up to 1000 Hz. Since this noise value is multiplied by the square root of the bandwidth, its contribution is low.

In the second region, the broadband noise of the amplifier is multiplied by the DC noise gain. Again, the average noise is multiplied by the square root of the bandwidth of that region. The contribution of noise in this region is also relatively low.

Chapter 10

The third, fourth and fifth regions are calculated in the same manner, with each region contributing more to the overall noise of the circuit. The sixth part of the noise equation in Figure 10.12 represents the noise contribution of the feedback resistor, R_2. The noise contribution of this resistor many or may not be significant depending on the magnitude of the resistor. This calculation will quickly demonstrate where the highest noise contribution is coming from and make it easier to refine the design.

Region e_1:
$$e_1 = (1 + R_2/R_1) * B \sqrt{(\ln(f_b/f_a))}$$

Region e_2:
$$e_2 = (1 + R_2/R_1) * e_n * \sqrt{(f_2 - f_1)}$$

Region e_3:
$$e_3 = (1 + R_2/R_1) * e_n * (1 \text{ Hz}/f_1) \sqrt{(f_2/3 - f_1/3)}$$

Region e_4:
$$e_4 = (1 + C_1/C_2) * en * \sqrt{(f_{AOL} - f_2)}$$

Region e_5:
$$e_5 = (1 + C_1/C_2) * en * \sqrt{(\pi/2(f_u - f_{AOL}))}$$

Region e_6:
$$e_6 = \sqrt{(4 * K * T * R_2 * (BW))}$$

Note: In this calculation, C_1 is the parallel combination of the input capacitors, or $C_{P-R1} \parallel 2C_{CM} \parallel C_{DIFF}$. C_2 is the parallel combination of the feedback capacitors or C_{P-R2}.

With all of this being said, SPICE is a useful tool as you verify your noise calculations.

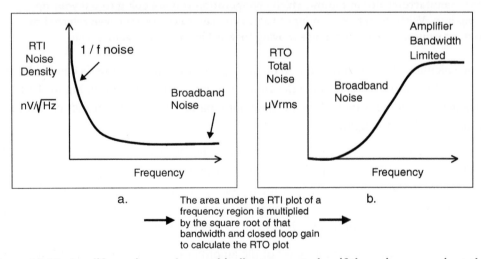

Figure 10.13: Amplifier noise can be graphically represented as if the noise source is at the input of the amplifier (a), which is otherwise known as referred-to-input (RTI), or as if it is at the output of the amplifier (b), which is otherwise known as referred-to-output (RTO).

The two graphs in Figure 10.13 demonstrate how SPICE can help you understand the noise in your circuit. The graph on the left (Figure 10.13a) shows the simulated noise response of an amplifier. The graph on the right (Figure 10.13b) provides the cumulative noise as frequency increases. You will notice that the noise is very low at the lower frequencies. This is because the lower bandwidths are multiplied by the square-root of a small number, the bandwidth. As frequency increases, the cumulative noise also increases. You would think that at higher frequencies the increases in noise would be less, because of the characteristics of the left-hand side graph (Figure 10.13a). As you can see, this is not true. The reason is that the bandwidth multiplier (square-root of the bandwidth) is larger at higher frequencies.

Going back to Figure 10.3 and 10.8, we concluded that reducing the resistor values was beneficial to a point. The next step is to reduce the amplifier noise. If the resistors are reduced 10x and the amplifiers are changed, the noise code width response in Figure 10.4 is reduced to twenty-one. This is not bad, considering we haven't changed the layout, just the devices.

A/D Converter Noise

Analog-to-digital converters are not well known for creating device noise but, like any other active device, they do. The most talked about noise from the A/D converter is quantization noise. Quantization noise is the noise that an A/D converter generates as a consequence of dividing the input signal into discrete "buckets." Figure 10.14 illustrates this. The width of these "buckets" is equal to the LSB size of the converter. The quantization noise of a converter determines the maximum signal-to-noise ratio ($SNR_{IDEAL} = 6.02n + 1.76dB$). This noise is immediately apparent in the converted signal. If you want more accuracy, you need to change to a converter with a higher number of bits. Just as a caveat, making this change does not guarantee a better SNR, because the converter may have other noise sources inside, but it is a good start.

Figure 10.14: The A/D converter does not convert the analog signal to an ideal value. Since the A/D converter has a discrete number of output conditions and the analog signal has an infinite number of voltage states, the A/D converter and error in the accuracy of the conversion. This is called the *quantization error*, which accounts for quantization noise.

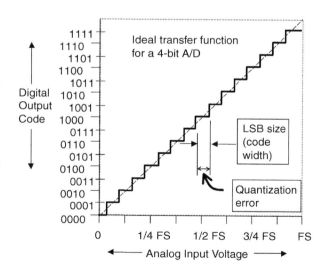

Chapter 10

There is also noise inside the A/D converter that comes from the internal transistors. This type of noise is discussed in Chapter 2 in more detail in the section "AC Specifications Imply Repeatablity." Just to recap the discussion in Chapter 2, AC domain specifications, such as signal-to-noise ratio (SNR), effective resolution (ER), signal-to-(noise + distortion) (SINAD), or effective number of bits (ENOB), help you understand how repeatable your A/D converter might be. These specifications do not imply accuracy, only repeatability.

Power Supply Noise

There are three fundamental types of power supply devices that you can use to deliver power to your circuit. They are the regulator (also low dropout regulator (LDO)), the switched power supply circuit (SPC), and the capacitive charge pumps. Use Figure 10.15 as an example of how you may want to connect these parts.

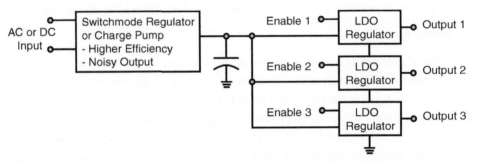

Figure 10.15: Power is taken from an AC (such as the wall) or DC (such as a battery) source and converted down (or up) to the required voltage level for the electronics. This conversion can be done in one stage or multiple stages, using LDO, switch-mode converters or capacitor charge pumps.

Low Dropout Regulators

Figure 10.16 shows a simplified CMOS LDO circuit. In this circuit, the input voltage directly supplies the power for the circuits, as well as the output load. Usually, the quiescent current of the LDO device is much lower than the output current.

Proper LDO operation requires that the input voltage is DC and higher in magnitude than the output voltage. When an input voltage is applied, the low-drift, bandgap reference voltage is established. The bandgap voltage and an operational amplifier are used to sense the resistor divider voltage at the output of the operational amplifier. This resistor divider establishes the output voltage magnitude. Regardless of input voltage, as long as it remains higher than $V_{DROPOUT} + V_{OUT}$, the output voltage will remain constant. The input source through the p-channel MOSFET, Q_1 provides the output current load. Q_1 can also be other types of transistors, like a Darlington pair of PNP bipolar transistors or an NPN transistor. The input current is equal to the output current plus the internal current required for bandgap generation, operational amplifier bias and p-channel MOSFET turn ON.

Figure 10.16: This figure illustrates a simplified diagram of an LDO, which uses a MOSFET, Q_1, to provide the output current drive. Q_1 is also the fundamental limiting factor at the output. In other topologies, Q_1 can be a bipolar transistor.

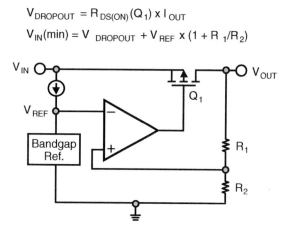

The LDO noise is relatively small as compared to active device's capability to reject that noise on its power supply pin. For instance, the spot noise of a typical LDO at 10 kHz is 1 µV/ √Hz rms. The noise rejection capability of a 10 MHz operational amplifier at that frequency is 70 dB. With this type of performance, the amplifier will reduce the noise from the LDO by 3162 times. Another way to look at this is that the LDO noise that gets into the amplifier (referred-to-input) is equal to approximately 316 pVrms.

In the general scheme of things, this is not critical. However, the noise rejection capability of the LDO is another story. If a SPC is in front of the LDO (as is the case with the circuit in Figure 10.3), it is highly probable that the LDO will not be able to remove that noise from the power supply line. In other words, LDOs are well known for "sharing" a lot of their noisy input with the rest of the circuit.

Switched Mode Power Supplies (SPC)

Figure 10.17 shows a simplified example of a Buck-SPC circuit. Chapter 9 covers the general operation of this circuit.

Figure 10.17: This diagram shows a Buck-SPC with the dynamic evaluation calculated for the device in the continuous mode. This power-supply requires an external inductor making manufacturing more difficult and is also capable of higher emitted noise.

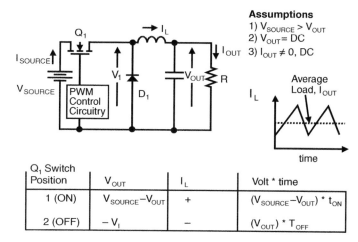

Chapter 10

Several issues have an impact on the amount of noise that this converter generates. The noise magnitude at the output of these types of devices depends on the type of switching mode employed. Generally, if the switched mode power converter is in a pulse width modulation (PWM) mode the peak-to-peak noise is lower than if the device is in a pulse frequency modulation (PFM) mode. The frequency of the switching noise of the converter is dependent of the converter's oscillator source. There is also an overshoot when a large amount of current is pulled from the device. Figure 10.18 shows an example of the switching noise from this type of device.

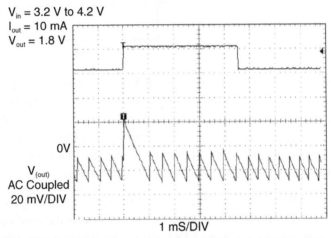

Figure 10.18: This figure shows the output voltage of a Buck-switch mode power supply converter with time. The converter is in its frequency mode of operation. The output signal produces a complex 20 mVp-p signal that has a switching frequency of 2000 Hz.

The device noise from a SPC changes into conducted noise once it is injected into the power supply trace. Figure 10.18 is not representative of the specific kind of noise that these types of devices generate. Every device is different, however there may be switched mode power supply noise in your circuit, and it may have a detrimental effect on the signal.

A second type of noise that comes from this type of device is radiated noise. The inductor in the circuit is a storage device and in that, it emits magnetic noise. The wall cube in Figure 10.5 has this type of power supply. The noise from the wall cube was effectively reduced with a ferrite bead (L_1).

Capacitive Charge Pump

Figure 10.19 shows a simplified circuit example of the charge pump, power-management system. The general operation of this circuit was discussed in detail in Chapter 9.

Power supply noise is generated at the frequency rate of the internal oscillator of this device. Again, the magnitude and frequency of the noise at the output of this device, which is injected

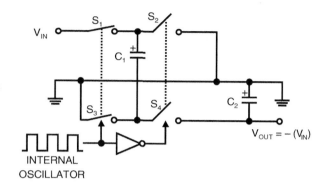

Figure 10.19: This figure shows an ideal switched capacitor charge-pump inverter. The low count in external components makes implementation of this circuit relatively easy.

into the power trace, is device dependent. For details, refer to your vendor's data sheet. This device also emits noise due to the switching action through the capacitors.

Minimizing Device Noise

If you are interested in minimizing the device noise is your circuit, you should first inspect the value of your resistors. If possible, make the resistors as low as possible. There are some design constraints that may prevent you from lowering the circuit resistors as low as you would like. For instance, amplifiers are generally not designed to drive low resistive loads. Another reason you may hesitate would be for power considerations. If you have resistors spanning the power supply, their power dissipation may be an issue, particularly in battery-powered applications. But with these constraints in mind, lowering the values of your resistors can provide a noise improvement of $\sqrt{F}$, where F is the reduction factor of the resistor.

Active device selection may also be the way to go, as you reduce the noise in your circuit. In this chapter, we discussed replacing the amplifiers in the circuit to lower noise amplifiers. This turned out to be critical, mainly because the amplifier noise (along with the resistor noise) was gained by the instrumentation amplifier configuration. Amplifier noise reduction is usually critical as the gain of the amplifier circuit increases. You may have noticed that I did not talk about the voltage reference in Figure 10.3. This is because that reference voltage and its noise goes straight through the instrumentation amplifier without the gain applied. But, if you are using a voltage reference that is gained or a reference to a converter with over 20-bits, the voltage reference selection may be critical.

Other active devices, such as the A/D converters or D/A converters should be selected to meet the requirements of your application. If you have a noisy device, you may have to switch it out for a higher-bit device.

Of the three power supply devices, the LDO has a relatively low noise output. However, that device passes noise almost directly from its input to output. This is the case with the LDO that is used in the circuit if Figure 10.3. The other two power supply devices do generate switching noise. If you sample this noise, you will find that it is not random like the resistor or operational amplifier. Instead, it has a complex signal that rides on top of the DC output of these devices. The magnitude of this small signal can be from a few millivolts to tens of millivolts. You can try to replace these devices with lower noise devices, but many times, you will have other requirements in your circuit, such as efficiency. If this is the case, a replacement device may not be feasible. But, there is help on the way as you read on. In the conducted noise section of this chapter we are going to talk about power supply filters for your active devices.

Conducted Noise

The third type of noise that affects the performance of analog devices is conductive noise. This type of noise already exists in the conductive paths of the circuit, such as the power lines or the signal path. Conducted noise mixes with the desired electrical signal. The best weapon against conducted noise is to go back to the source and implement noise reduction strategies on the offending device or radiating source. But you are going to find that sometimes you know the origin of the conducted noise and you can't further reduce it at the source. For instance, you may need to use a switching power supply because of its improved efficiency as compared to an LDO. This power supply device allows you to run your application circuit with lower power dissipation. Now that this noise problem has changed from device noise to conducted noise, there are filtering techniques that you can use to overcome this noise problem. The origin of conducted noise is either device noise or radiated noise.

Noise in the Signal Path

Signal path noise can come from a variety of devices. For instance, resistor noise is a likely candidate in your circuit. Amplifier noise is another. Earlier in this chapter, we discussed the techniques that you might employ to reduce noise at the source. If you can't change the components or there are no lower-noise alternatives, the next tactic is to insert some kind of filter. Your fundamental choices are low-pass, high-pass or bandpass. Of these three types of filters, you will find the most beneficial one is the low-pass filter.

You will note that the circuit in Figure 10.3 does not have an anti-aliasing filter. As the data shows, this oversight has caused noise problems in the circuit. When the board has a 2nd order, 10 Hz, anti-aliasing filter inserted between the output of the instrumentation amplifier and the input of the A/D converter, the conversion response improves dramatically (see Figure 10.5).

Analog filtering can remove noise superimposed on the analog signal before it reaches the A/D converter. In particular, this includes extraneous noise peaks. Analog-to-digital converters will convert the signal that is present on its input. This signal could include that sensor voltage signal or noise. The anti-aliasing filter removes the higher frequency noise from the conversion process. Chapter 4 covers the topic of low-pass filters in detail.

Noise in Your Power Supply Bus

There are several things you can do to reduce noise in your power supply. The most commonly recommended strategy is to have a bypass or decoupling capacitor straddling the supply pin to ground of every active device. You will find with analog devices, the text of the product data sheets recommends bypass capacitor values.

Analog devices and digital devices all require these types of capacitors. In both cases, these devices require a capacitor as close to the power supply pin(s) with a common value for this capacitor of 0.1 µF. A second class of capacitor in the system is required at the power supply source. The value of this capacitor is usually about 10 µF.

Bypass capacitors belong in two locations on the board: one at the power supply (10 µF to 100 µF or both) and one for every active device (digital and analog). The value of the device's bypass capacitor is dependent on the device in question. If the bandwidth of the device is less than or equal to ~1 MHz, a 1 µF will reduce injected noise dramatically. If the bandwidth of the device is above ~10 MHz, a 0.1 µF capacitor is probably appropriate. In between these two frequencies, both or either one could be used. Refer to the manufacturer's guidelines for specifics.

Every active device on the board requires a bypass capacitor. It must be placed as close as possible to the power supply pin of the device as shown in Figure 10.20. If you have two bypass capacitors for one device, the smaller one should be closest to the device pin. Finally, the lead length of the bypass capacitor should be as short as possible.

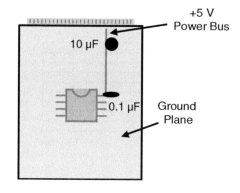

Figure 10.20: If you want to reduce the overall noise in your circuit (as well as enhance stability) the bypass or decoupling capacitors shown in this figure are crucial. The smaller value capacitor (0.1 µF) is as close to the device power pin as possible. The higher value capacitor (10 µF) is as close to the power supply source as possible.

Bypass or decoupling capacitors and their placement on the board are just common sense for both types of devices, but interesting enough, for different reasons. In the analog layout design, bypass capacitors generally serve the purpose of redirecting high frequency signals on the power supply that would otherwise enter into the sensitive analog chip through the power supply pin. Generally speaking, these high frequency signals occur at frequencies beyond the analog device's capability to reject those signals. The possible consequences of not using a bypass capacitor in your analog circuit results in the addition of undue noise to the signal path and worse yet, oscillation.

Chapter 10

For digital devices, such as controllers and processors, decoupling capacitors are required, but for a different reason. One of the functions of these capacitors serves as a "mini" charge reservoir. Frequently in digital circuits, a great deal of current is required to execute the transitions of the changing gates. Because of the switching transient currents that occur on the chip and throughout the circuit board, having additional charge "on call" is advantageous. The consequence of not having enough charge locally to execute this switching action could result in a significant change in the power supply voltage. When the voltage change is too large, it will cause the digital signal level to go into the indeterminate state, more than likely resulting in erroneous operation of the state machines in the digital device. The switching current passing through the circuit board traces would cause this change in voltage. The circuit board traces have parasitic inductance, and you can calculate the change in voltage results by using the formula:

$$V = L\delta I/\delta t$$

Where V = voltage change
 L = board trace inductance
 δI = change in current through the trace
 δt = the time it takes for the current to change

So for multiple reasons, it is a good idea to bypass (or decouple) the power supply at the power supply and at the power supply pin of active devices.

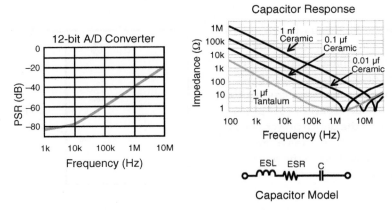

Figure 10.21: Bypass capacitors reduce the noise on the power supply pin of a device at frequencies where the power supply rejection capability of the device is too low to reject noise.

Although, the manufacturer usually recommends a bypass capacitor value, you can determine the impact of changing the recommended value. The plot on the right (a) in Figure 10.21 shows the power supply rejection capability of a 12-bit A/D converter. If power supply noise exists at lower frequencies, the converter will attenuate them by ~80 dB or 10,000 times. But, at higher frequencies, the converter is less able to reject signals on the supply.

Since this is a 12-bit converter, any signal injection that causes less than a ¼ LSB error is not noticed. But as the interfering signal from the power supply starts to cause conversion errors, attenuation of that signal is required. In this example, the power supply to the converter is 5 V and it has ± 20 mV noise riding on it. Near DC, this noise is attenuated 10,000 times by the converter, or to a voltage level of 2 µV peak-to-peak. You would never see this noise at the output of the converter. However, the point where the noise is equal to about ± ¼ LSB or ± ¼ × (FSR / 2^{12}) or ± 0.31 mV. The required attenuation for the ± 20 mV noise signal is −36.3 dB or lower. This starts to occur at approximately 2 MHz as you go up in frequency. If the noise signal is not attenuated at frequency of 2 MHz or higher, the noise will start to make its way into the output code.

An easy solution to this problem is to select a bypass capacitor that passes higher frequency signals to ground. The graph on the left-hand side of Figure 10.21 shows the frequency response of several capacitors. The frequencies where these curves extend down towards zero are the frequencies that are passed to the ground plane. For the application circuit, where the 12-bit A/D converter of Figure 10.21 is used, the best bypass capacitor would be a 0.1 µF ceramic.

Sometimes the simple bypass capacitor is not adequate. For instance, in the circuit in Figure 10.3 there was a power-supply small-signal that came from the wall cube that appeared at the output of the A/D converter. Figure 10.22 shows this signal in the time domain.

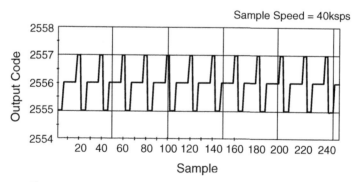

Figure 10.22: Power supply noise from the wall cube in Figure 10.3. The power supply frequency that appears at the output of the A/D Converter is approximately 2 kHz.

You cannot filter this low-frequency noise with a bypass capacitor, unless you are willing to use an extraordinary high valued, large, expensive capacitor. This is not recommended. There are other techniques that you can use to filter power supply noise of this sort.

The most common power supply filtering for analog circuits uses a bypass or decouple capacitor (Figure 10.23a). This will reject frequencies that are present on the power supply-line when the analog device is not able to reject an AC signal with its power supply rejection

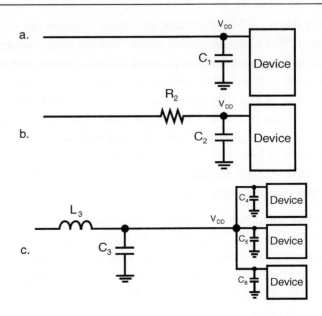

Figure 10.23: You can use these three power supply filters for higher frequency filtering (a), as a voltage divider, or (b), a ferrite bead.

capability. If lower frequency noise is still injected into the analog signal path (as illustrated in Figure 10.22), an additional component may be required is series with the device as shown in Figure 10.23b and 9.21c. The resistor in series with the device that is shown in Figure 10.23b creates a low-pass filter, in conjunction with C_2. This low-pass filter can be designed to eliminate lower frequency noise if the correct resistor/ capacitor combination is chosen. This type of low-pass filter begins to attenuate power supply noise at the frequency, $1/(2\pi RC)$.

The negative side of using the strategy is the lost of power supply voltage near DC. This loss is due to the voltage drop across R_2. Consequently, if you are trying to design a low frequency filter, high value capacitors are required. This can become a vicious circle.

Figure 10.23.c shows an alternative circuit. In this circuit, a ferrite bead is in series with the device. For this type of circuit, the ferrite bead can accept DC current, while rejecting low frequency AC signals. At low frequencies, ferrite beads are inductive. This combines with the bypass capacitor for form a low-pass LC filter. At higher frequencies the impedance of these beads if primarily resistive. Since the ferrite bead is an inductor that carries high currents, the DC loss across this element is very low.

Proper selection of your ferrite bead can reduce power supply noise. For example a switcher that is operating with a 40 mVp-p ripple at ~150 kHz down to 3 mVp-p. This is can be achieved with a 50 µH ferrite bead and a 100 µF tantalum capacitor (C_3).

Revisit Our Application Circuit with these Low-Noise Improvements

I always like to bring us back to reality (or better yet, the lab) to prove that things work the way we are theorizing they will. So let's return to the circuit in Figure 10.5. I am showing this circuit again in Figure 10.24 so you won't have to leaf through the book pages.

Noise – The Three Categories: Device, Conducted and Emitted

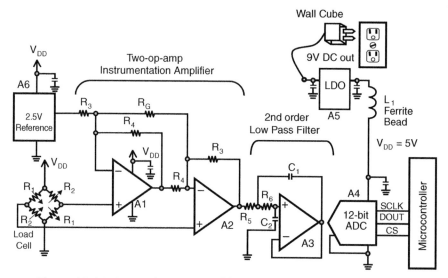

Figure 10.24: From Figure 10.3, this circuit has reduced resistors, lower noise amplifiers, a 2nd order low-pass filter, and an inductive choke or ferrite bead on the power supply. You can implement all of this tactics in the circuit to reduce analog noise and achieve 12-bit accuracy from the 12-bit ADC.

In this new and improved circuit, I reduced the resistors in the instrumentation amplifier by 10×. I then swapped out the amplifiers for lower noise device. As I moved on to conducted noise problems, I added a 2nd order, low-pass filter in the signal path because I know that my instrumentation amplifier circuit had a high gain. Then I added the appropriate bypass capacitors near the active devices. From there, I put in my ground plane. This addition made a big difference because the ground trace resistance was reduced and the board now had some shielding from outside interference. Finally, I added a ferrite bead to the power supply trace to remove any lower frequency noise from the power supply. These noise reduction activities changed my "~7-bit" converter to a true 12-bit converter.

Chapter 10 References

Noise Reduction Techniques in Electronic Systems, Ott, Henry W., John Wiley & Sons, NY, 1976.

Noise and Other Interfering Signals, Morrison, Ralph, John Wiley & Sons, NY, 1991.

Statistics for Experimenters, Box, Hunter, John Wiley & Sons, Inc., 1978.

"Noise and Operational Amplifier Circuits," Smith, Sheingold, Analog Dialogue 3-1, 1969.

"D-C Amplifier Noise Revisited," Ryan, Scranton, Analog Dialogue 18-1, 1984.

"Understanding Data Converters Frequency Domain Specifications," AN-4, Datel.

"Comparison of Noise Performance Between a Fet Transimpedance Amplifier and a Switched Integrator," Baker, Bonnie C., AN-057, Texas Instruments.

"The Effect of Direct Current on the Inductance of a Ferrite Core," Fair-Rite Products Corp.

CHAPTER 11

Layout/Grounding (Precision, High Speed and Digital)

CHAPTER 11

Layout/Grounding (Precision, High Speed and Digital)

The ratio of digital designers versus the analog designers is increasing. This is not a news flash, at least if you have read the first chapter of this book. Although the emphasis on digital design is providing significant advances in electronic, end products, there is still and will always be a portion of circuit design that interfaces with the analog or real-world. Could it be possible that analog layout differ from digital layout techniques? There is some similarity in layout strategies between these two domains. The differences can make an easy circuit layout design less than optimum if you are trying to achieve good results. In this chapter, we will discuss five topics. The first topic covers fundamental similarities and differences between analog and digital layout. Then we will talk about the hidden components (resistors, inductors, and capacitors) embedded in your PC board. The next section of this chapter will talk about how to improve your A/D converter accuracy and resolution. Getting another converter will not help here. Focusing on the interaction between the PCB and your converter will improve your results. This will be followed by the fourth section where we will discuss two-layer layout techniques. Finally, we will end with an example of how to do a poor layout and then how to fix it.

The Similarities of Analog and Digital Layout Practices

There are many similarities between analog and digital layout practices. As the digital systems get faster and faster, the digital circuit looks more analog than not. When you talk about similarities between these domains, the use of bypass capacitors and power plane designs are basically the same. Differences pop-up when you talk about switching noise and the location of devices on the board.

Bypass or Decoupling Capacitors

In terms of layout, analog devices and digital devices all require these types of capacitors. Both types of devices require that you position one capacitor as close to the power supply pin(s) as possible. A common value for this capacitor is 0.1 µF, but it is not unusual to find a 1 µF bypass capacitor (for lower frequency circuits) or a 0.01 µF capacitor in higher frequency circuits. The selection of the proper capacitor for your circuit is discussed in detail in Chapter 6. A second class of bypass or decoupling capacitor in the system is required at the power supply source. The value of this capacitor is usually about 10 µF.

Chapter 11

Figure 11.1 shows the position of these capacitors. The values of these capacitors can vary by being ten times higher or lower, but they are both required to have short leads. The inductance of shorter leads is smaller, reducing the chances of having a "tank" circuit. The smaller value capacitor should be as close to the device as possible and the higher value capacitor should be as close to the power supply source as possible.

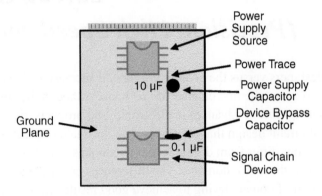

Figure 11.1: In analog and digital PCB design, you should place the bypass or decouple capacitors (0.1 µF) as close to the device as possible. You should also place the power supply decoupling-capacitor (10 µF) at the power-source or where the power-bus enters the board. In all cases, these capacitors should have short leads.

The placement of the bypass or decoupling capacitors are just common sense for both types of designs, but interesting enough, for different reasons. In the analog layout design, bypass capacitors generally serve the purpose of redirecting high frequency signals on the power supply trace. This noise would otherwise enter into the sensitive analog chip, through the power supply pin. Generally, these high frequency signals occur at frequencies beyond rejection capability of the analog device. The possible consequences of not using a bypass capacitor in your analog circuit results in the addition of undue noise to the signal path or worse yet, oscillation.

For digital devices, such as controllers and processors, the decoupling capacitor on the power supply pin are required, but for a different reason. One of the functions of these capacitors serves as a "mini" charge reservoir. Frequently in digital circuits, a great deal of current is required to execute the transitions of the changing gate states. Because of the switching transient currents that occur on the chip and throughout the circuit board, having additional charge "on-call" is advantageous. The consequence of not having enough charge locally to execute this switching action could result in a significant dynamic and static change in the power supply voltage. When the voltage change is too large, it will cause the digital signal level to go into the indeterminate state. But more than likely, the state machines in the digital device will operate erroneously. The switching current passing through the circuit board

Layout/Grounding (Precision, High Speed and Digital)

traces cause this change in voltage. The circuit board traces have parasitic inductance. You can calculate the change in voltage results with this formula:

$V = L\delta I/\delta t$

where V = voltage change
 L = board trace inductance
 δI = change in current through the trace
 δt = the time it takes for the current to change

So for multiple reasons, it is a good idea to bypass (or decouple) the power supply at the power supply and at the power supply pin of all of the active devices.

The Power and Ground Should Be Routed Together

When you match power and ground traces with respect to location, you lesson the opportunities for EMI. If you don't match power and ground, system loops are part of the layout. The possibility of seeing "noisy " results without explanation is real. Figure 11.2 shows an example of a PCB design with the unmatched power and ground traces.

The loop area in Figure 11.2 is 697 cm². This loop is a perfect antenna for noise in the area. With this board, you may be able to pick up radio signals. In the 1980s, one of the German engineers that I worked with was able to design boards of this class and "pick-up" radio-free Europe.

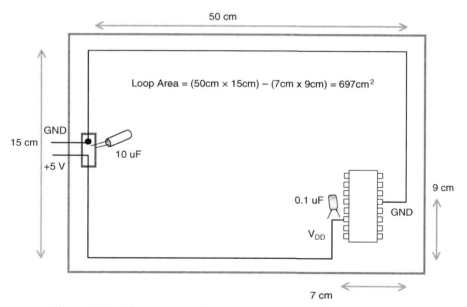

Figure 11.2: The power and ground traces are laid out using different routes to the device on this board. This mismatch opens the opportunity for EMI into the electronics of this board.

Chapter 11

Figure 11.3 shows a dramatic decrease in radiated noise off the board for induced voltages in the loop. This is because there is a decrease of radiated noise off the board and around the board.

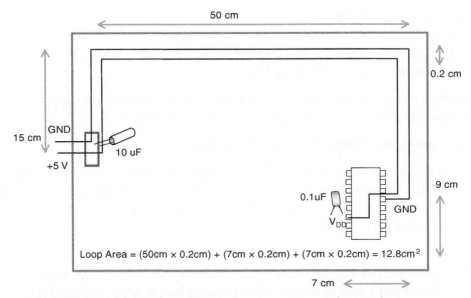

Figure 11.3: In this one layer board, the power trace and ground trace are laid next to each other on their way to the device on this board. Figure 11.2 shows a board where the traces are better matched. The opportunity for EMI into the electronics of this board is lessened by 679/12.8 or ~54x.

In Figure 11.3, the signal and ground line are next to each other. This greatly reduces the loop area. An even better solution would be to have a ground plane, which would be underneath the power supply trace. An even better solution would be to have a ground plane and a separate power plane.

Where the Domains Differ – Ground Planes Can Be a Problem

The fundamentals of circuit board layout apply to analog circuits as well as digital circuits. One fundamental rule of thumb is to use uninterrupted ground planes. This common practice reduces the effects of δI/δt (change in current with time) in digital circuits. In digital circuits, the change in current with time changes the potential of ground. In analog circuits, injected noise is caused by δI/δt. But, when comparing digital and analog circuits, you should exercise an added precaution with analog circuits in order to keep the digital signal lines and return paths in the ground plane as far away from the analog circuitry as possible. This can be done by connecting the analog ground plane separately to the system ground connect or having the analog circuitry at the farthest side of the board—that is, at the end of the line. This is done so that signal paths have a minimal amount of interference from external sources. The opposite is not true for digital circuitry. The digital circuitry can tolerate a great deal of noise on the ground plane before problems start to appear.

Location of Components

In every PCB design, you should separate the noisy and quiet portions of the circuit, as mentioned above. Generally, the digital circuitry is "rich" with noise. Alternatively, digital circuitry is less sensitive to this type of noise because of the larger voltage noise margins. When you look at analog circuits, you will easily find that they are not as forgiving as the digital circuits. The voltage noise margins of the analog circuitry are much smaller. Of the two domains, the analog domain is most sensitive to switching noise. In the layout of a mixed signal system, you should separate the two domains. Figure 11.4 shows this is graphically.

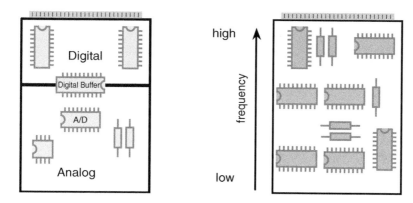

Figure 11.4: If possible, (a) the digital and analog portion of circuits should be separated in order to separate the digital switching activity from the analog circuitry. Additionally, (b) the high frequency should be separated from the low frequency where possible, keeping the higher frequency components closer to the board connector.

The general rules of thumb are to keep the analog and digital portions of the circuit separate, with the digital circuitry closest to the connector. This is done so that the fast changing digital signals never "go past" the analog chips. A second general guideline is to place the higher frequency devices closer to the connector than the lower frequency devices. In this case, higher frequency noise will not inject into the lower frequency devices.

Where the Board and Component Parasitics Can Do the Most Damage

The major classes of parasitics generated by the PC board layout come in the form of resistors, capacitors, and inductors. For instance, you can build PCB resistors with your traces that

Chapter 11

span between components. You can build unintentional capacitors into the board with traces, soldering pads, and parallel traces. Unintentional inductors come from loop inductance, mutual inductance, and vias. All of these parasitics stand a chance of interfering with the effectiveness of your circuit as you transition from the circuit diagram to the actual PCB. You will clearly see in this section of the chapter the most troublesome class of board parasitics and see examples of where these parasitics the effect on circuit performance.

Feeling the Pain of Those Unnecessary Capacitors

In Chapter 9, we discussed how you could inadvertently build capacitors into your board. To quickly review this concept, you can design a capacitor into a board by simply placing two traces close to each other. This can be done by placing the two traces, on top of the other with two layers (which is harder to see) or by placing them beside each other on the same layer. Usually, you would build layout capacitors by placing two parallel traces close together. The formulas in Figure 11.5 show how you can calculate the value of this type of capacitor.

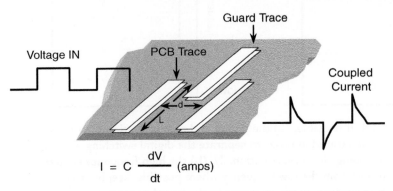

Figure 11.5: You can easily place capacitors into a PCB by laying out two traces in close proximity. With this type of capacitor, fast voltage changes on one trace can initiate a current signal in the other trace. (Chapter 9 also discusses this issue.)

In both trace configurations, changes in voltage with time ($\delta V/\delta t$) on one trace could generate a current on a second trace. If the second trace is high impedance, the e-field creates current, which converts to voltage. Typically, you will find fast voltage transients on the digital side of the mixed signal design. If the traces that have these fast, voltage transients are in close proximity of high impedance, analog traces, this type of error will be very disruptive with analog circuitry accuracy. Analog circuitry has two strikes against it in this environment. The noise margins are much lower than digital and it is not unusual to have high impedance traces.

Layout/Grounding (Precision, High Speed and Digital)

You can easily minimize this type of phenomena using one of two techniques. The most commonly used technique is to change the dimensions between the traces as the capacitor equation suggests. The most effect dimension to change is the distance between the two offending traces. It should be noted that the variable, "d", is in the denominator of the capacitor equation. As "d" is increased, the capacitance will decrease. The length of the two traces is another variable that you can change. In this case, if you reduce the length ("L"), this also reduces the capacitance between the two traces.

Another technique used is to lay a ground trace between the two offending traces. Not only is the ground trace low impedance, but an additional trace like this will break up the E-fields that are causing the disturbance.

This type of capacitor can cause problems in mixed signal circuits where sensitive, high impedance, analog traces are in close proximity to digital traces. For example, the circuit in Figure 11.6 has the potential to have this type of problem.

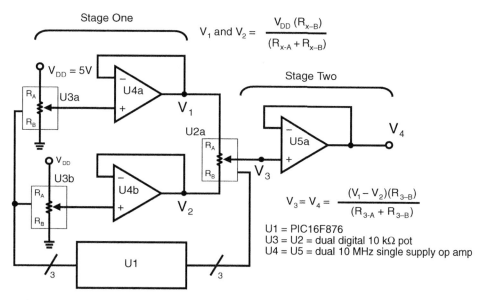

Figure 11.6: You can build a 16-bit DAC using three 8-bit digital potentiometers and three amplifiers to provide 65,536 different output voltages. If V_{DD} is 5 V in this system the resolution or LSB size of this DAC is 76.3 µV.

To quickly explain the circuit operation in Figure 11.6, a 16-bit DAC uses three 8-bit digital potentiometers and three CMOS operational amplifiers. To the left side of this figure, two digital potentiometers (U3a and U3b) span across V_{DD} to ground with the wiper output connected to the non-inverting input of two amplifiers (U4a and U4b). You program the digital potentiometers, U2 and U3 by using an SPI interface between the microcontroller, U1. In this configuration, each digital potentiometer operates as an 8-bit multiplying DAC. If V_{DD} is equal to 5 V, the LSB size of these DACs is equal to 19.61 mV.

Chapter 11

In this circuit, you connect the wipers digital potentiometers (U4a, U4b) to the non-inverting inputs of two buffer amplifiers. In this configuration, the inputs to the amplifiers are high impedance, which isolates the digital potentiometers from the rest of the circuit. The output swing restrictions of the second stage of this amplifier configuration is not violated.

To have this circuit perform as a 16-bit DAC (U2$_a$), a third digital potentiometer spans across the output of these two amplifiers, U4a, and U4b. The programmed setting of U3a and U3b sets the voltage across the digital potentiometer. Again, if V$_{DD}$ is 5 V it is possible to program the output of U3a and U3b 19. 61 mV apart. With this size of voltage across the third 8-bit digital potentiometer (R$_3$), the LSB size of this circuit from left to right is 76.3 µV. Table 11.1 shows the critical device specifications that give optimum performance with this circuit.

Table 11.1: From the long list of specifications that each of the devices have, there are a handful of key specifications that make this circuit more successful when it is used to provide DC reference voltages or arbitrary wave forms.

Device	Specification		Purpose
Digital Potentiometers	Number of bits	8-bits	Determines the overall LSB size and resolution of the circuit.
	Nominal resistance (resistive element)	10 kΩ (typ)	The lower this resistance is the lower the noise contribution will be to the overall circuit. The trade off is that the current consumption of the circuit is high with these lower resistances.
	DNL	± 1LSB (max)	Good Differential Non-Linearity is needed to insure no missing codes occur in this circuit which allows for a possible 16-bit operation.
	Voltage Noise Density (for half of the resistive element)	9 nV/√Hz @ 1 kHz (typ)	If the noise contribution of these devices is too high it will take away from the ability to get 16-bit noise free performance. Selecting lower resistive elements can reduce the digital potentiometer noise.
Operational Amplifiers	Input Bias Current, I$_B$	1 pA @ 25°C (max)	Higher I$_B$ will cause a DC error across the potentiometer. CMOS amplifiers were chosen for this circuit for that reason.
	Input Offset Voltage	500 µV (max)	A difference in amplifier offset error between U4a and U4b could compromise the DNL of the overall system.
	Voltage Noise Density	8.7 nV/√Hz @10kHz (typ)	If the noise contribution of these devices is too high it will take away from the ability to get 16-bit accurate performance. Selecting lower noise amplifiers can reduce amplifier noise.

Layout/Grounding (Precision, High Speed and Digital)

You can use this circuit in two basic modes of operation. The first mode would be if you wanted a programmable, adjustable, DC reference. In this mode, you only use the digital portion of the circuit occasionally and certainly not during normal operation. The second mode would be if you used the circuit as an arbitrary wave generator. In this mode, the digital portion of the circuit is an intimate part of the circuit operation. In this mode, the risk of capacitive coupling may occur.

Figure 11.7 shows the first pass layout of the circuit. You can quickly design this circuit in your lab without attention to detail. The consequences of placing digital traces next to high impedance, analog lines were overlooked in the layout review. This speaks strongly to doing it right the first time, but to your benefit, I made this mistake and you can see how I made significant improvements.

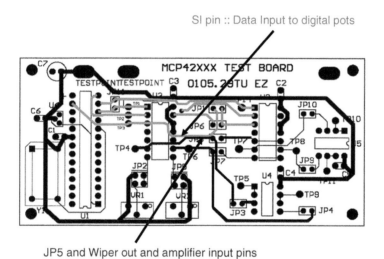

Figure 11.7: This is the first attempt at the layout for the circuit in Figure 11.6. In this figure, you can see that a critical, high impedance, analog line is very close to a digital trace. This configuration produces inconsistent noise on the analog line because the data input code on that particular digital trace changes. These changes are dependent on the programming requirements for the digital potentiometer.

If you take a look at this layout, it is obvious where a potential problem is. The arrow is pointing to an analog trace. This trace from the wiper of U3a to the high impedance amplifier input of U4a. The digital trace that is pointed out carries the digital word that programs the digital potentiometer settings.

On the bench, I measured the digital signal that was coupled into the sensitive, analog wiper, trace. Figure 11.8 shows the scope photo.

Chapter 11

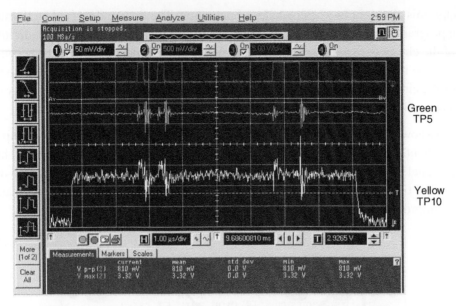

Figure 11.8: In this scope photo, the top trace was taken at JP1 (digital word to the digital potentiometers), the second trace on JP5 (noise on the adjacent analog trace), and the bottom yellow trace is taken at TP10 (noise at the output of the 16-bit DAC).

The digital signal that is programming the digital potentiometers in the system has transmitted from trace to trace onto an analog line that is being held at a DC voltage. This noise propagates through the analog portion of the circuit all the way out to the third, digital potentiometer (U5a). The third digital potentiometer is toggling between two output states.

What is the solution to this problem? Basically, you should separate the traces. Figure 11.9 shows an improved layout solution.

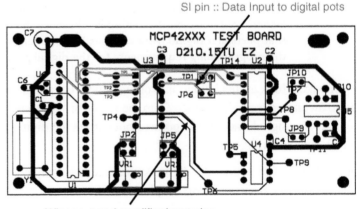

Figure 11.9: With a new layout, the analog lines are separate from the digital lines. This distance has eliminated the digital noise that was causing interference in the previous layout.

Layout/Grounding (Precision, High Speed and Digital)

Figure 11.10 shows the results of the layout change. With the analog and digital traces carefully kept apart, this circuit becomes a very clean 16-bit DAC. This trace shows a single code transition of the third digital potentiometer of 76.29 μV. You may notice that the oscilloscope scale is 80 mV/div and that the amplitude of this code change is approximately 80 mV. In the lab, the equipment forced us to gain the output of the 16-bit DAC by 1000x.

Once again, when the digital and analog domains meet, careful layout is critical if you intend to have a successful final PCB implementation. In particular, active digital traces close to high impedance, analog traces will cause serious coupling noise. You can avoid this noise coupling phenomena by putting distance between traces.

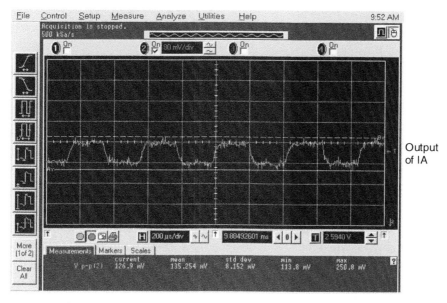

Figure 11.10: The 16-bit DAC in this new layout is showing a single code transition with no digital noise from the communication to the digital potentiometers.

Inductors Designed into the PCB

The way that an inductor is designed into a board is similar to the construction of a capacitor. Again this is done by placing two traces, one on top of the other with two layers or by placing them beside each other on the same layer, as shown in Figure 11.11. In both trace configurations, changes in current with time ($\delta I/\delta t$) on one trace could generate a voltage in the same trace due to the inductance on that trace and initiate a proportional current on the second trace due to the mutual inductance. If the voltage change is high enough on the primary trace, the disturbance can reduce the voltage margin of the digital circuitry enough to cause errors. This phenomenon is not necessarily reserved for digital circuits, but more common in that environment because of the larger, seemingly instantaneous switching currents.

Chapter 11

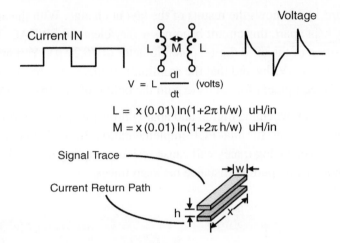

Figure 11.11: If you pay little attention to the placement of traces, you can create line and mutual inductance with the traces in a PCB. This kind of parasitic element is most detrimental to the circuit operation where digital switching circuits reside.

To eliminate potential noise for EMI sources it is best to separate quiet analog lines versus noisy I/O ports. Try to implement low impedance power and ground networks, minimize inductance in conductors for digital circuits and minimize capacitive coupling in analog circuits.

Layout Techniques That Improve ADC Accuracy and Resolution

Initially, analog-to-digital (A/D) converters rose from an analog paradigm where a large percentage of the physical silicon was analog. As the progression of new design topologies evolves, this paradigm is shifting to where slower speed A/D converters are predominately digital. Even with this on-chip shift from analog to digital, the PCB layout practices have not changed. Now as always, when the layout designer is working with mixed signal circuits, you still need key layout knowledge in order to implement an effective layout. This section of the chapter will look at the PCB layout strategies required for A/D converters using successive approximation register (SAR) and sigma-delta topologies.

SAR Converter Layout

SAR A/D converters can be found with 8-bit, 10-bit, 12-bit, 16-bit and sometimes 18-bit resolution. Originally, the process and architecture for these converters was bipolar with R-2R ladders. But recently these devices have migrated to a CMOS process with a capacitive charge distribution topology. Needless to say, the system layout strategy for these converters has not changed with this migration. The basic approach to layout is consistent, except for higher resolution devices. These devices require more attention to the prevention of digital feedback from the serial or parallel output interface of the converter.

The SAR converter is predominately analog in terms of circuitry and the amount of real estate dedicated to the different domains on the chip. Figure 11.12 shows a block diagram of a 12-bit, CMOS SAR converter.

Layout/Grounding (Precision, High Speed and Digital)

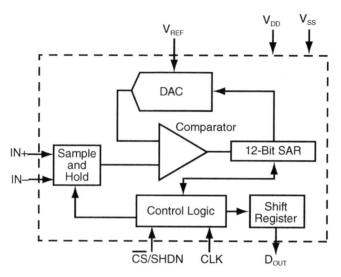

Figure 11.12: This is a block diagram of a 12-bit CMOS SAR A/D converter. This converter uses a charge distribution across a capacitive array. Refer to Chapter 2 for more information on how these devices actually work.

These types of converters can have several pins for the ground and power connections. The pin names are often misleading in that you cannot differentiate between the analog and digital connections with the pin label. These labels do not necessarily describe the system connections to the PCB, but rather they identify how the digital and analog currents come off the chip. Knowing this information and understanding that the primary real estate consumed on the SAR converter chip is analog, it makes sense to connect the power and ground pins on the same planes. And since the converter is primarily an analog chip, placing the pins of the device on the analog planes is very appropriate.

Figure 11.13 shows the pinout for a representative sample of 10-bit and 12-bit converters.

Figure 11.13: The SAR converter, regardless of resolution, usually has at least two ground connects; AGND (V_{SS}) and DGND (V_{SS}). The converters illustrated here are the MCP3008 and MCP3201 from Microchip.

275

Chapter 11

With these devices, the ground signal is usually directed off the chip with two pins; AGND and DGND. The power is applied to a single pin. When implementing the PCB layout, you should connect AGND and DGND to the analog ground plane. The analog and digital power pins should also be connected to the analog power plane or at least connected to the analog power train with proper bypass capacitors as close to each pin as possible. The only reason that these devices would have only one ground pin and one positive supply pin is due to package pin limitations. However, separate grounds on the chip enhance the probability of getting good and repeatable accuracy from the converter.

With all of the converters, the power supply strategy should be to connect all grounds, positive supply and negative supply pins to the analog planes. In addition, you should connect the COM pin or IN- pin associated with the input signal as close to the signal ground as possible.

Higher resolution SAR converters (16- and 18-bit converters) require a little more consideration in terms of separating the digital noise from the quiet analog converter and power planes. When you interface these devices to a microcontroller, external digital buffers should be used in order to achieve clean operation. Although, these types of SAR converters typically have internal double buffers at the digital output, you can use external buffers to further isolate the digital bus noise from the analog circuitry in the converter. Figure 11.14 shows an appropriate power strategy for this type of system.

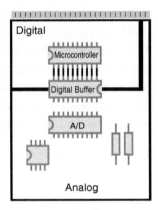

Figure 11.14: With high-resolution SAR A/D converters, you should connect the analog planes to the converter power and ground. You should then buffer the digital output of the A/D converter by using external 3-state output buffers. These buffers provide isolation between the analog and digital side, as well as high-drive capability.

Precision Sigma-Delta Layout Strategies

The silicon area of the precision sigma-delta A/D converter is predominately digital. In the early days, when this type of converter was first being produced, this shift in the paradigm prompted users to separate the digital noise from the analog noise by using the PCB planes. As with the SAR A/D converter, these types of A/D converters can have multiple analog- and digital-ground and power pins. Once again, the common tendency of a digital or analog design engineer is to try separating these pins into separate planes. Unfortunately, this is a misguided tendency, particularly if you intend to solve critical noise problems with the 16-bit to 24-bit accuracy devices.

Layout/Grounding (Precision, High Speed and Digital)

With a high-resolution sigma-delta converters that has a 10 Hz data rate, the clock (internal or external) to the converter could be as high as 10 MHz or 20 MHz. You would use this high frequency clock for switching the modulator and running the oversampling engine. With these circuits, you should connect the AGND and DGND pins together on the same ground plane, as is the case with the SAR converter. Additionally, you should connect the analog and digital power pins together, preferably on the same plane. The requirements on the analog and digital power planes are the same as with the high-resolution SAR converters.

A ground plane is mandatory, which implies that at a minimum you need a two-layer board. On this double-sided board, the ground plane should cover at least 75% of the area if not more. You should keep interruptions in the plane to an absolute minimum. The purpose of this ground plane layer is to reduce grounding resistance and inductance as well as provide a shield against electro-magnetic interference (EMI) and radio-frequency interference (RFI). If the circuit, interconnect traces need to be on the ground-plane side of the board, they should be as short as possible and perpendicular to the ground current return paths.

You can get away without separating the analog and digital pins of low precision A/D converters, such as 6-, 8-, or maybe even 10-bit converters. But as the resolution/accuracy increases with your converter selection, the layout requirements also become more stringent. In both cases, with high-resolution SAR A/D converters and sigma-delta converters you need to connect them directly to the lower noise, analog ground and power planes.

The Art of Laying Out Two-Layer Boards

In this highly competitive marketplace, the cost objective usually dictates that a designer use two-layer boards in the design. Although the multi-layer board (4-, 6-, and 8-layers) allows the designer to build cleaner solutions in terms of size, noise, and performance, financial pressures force the engineer to rethink layout strategies with the two-layer board in mind. In this section of this chapter we will discuss the use or misuse of auto routing, the concept of current return paths with and without ground planes, and recommendations for component placement where two layer boards are concerned.

Pay Now or Pay Later with the Auto Router and Analog Circuits

It is tempting to use the auto router when designing printed circuit board (PCB). More often than not, a purely digital board, (especially if the signals are relatively slow, and the circuit density is low) will work just fine. But as you try to layout analog, mixed signal or high-speed circuits with the auto routing tool that is available with your layout software, there may be some issues. The probability of creating serious circuit performance problems is very real.

Figure 11.15 shows the auto routed top layer of a two-layer board. The bottom layer of this board is in Figure 11.15 and 11.16 and the circuit diagram for these layout layers is in Figure 11.17 and Figure 11.18.

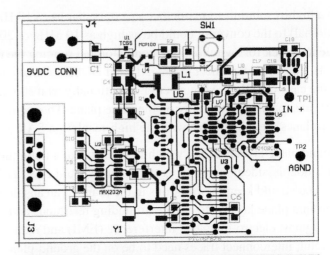

Figure 11.15: Top layer of an auto-routed layout of circuit diagram shown in Figure 11.17 and 11.18.

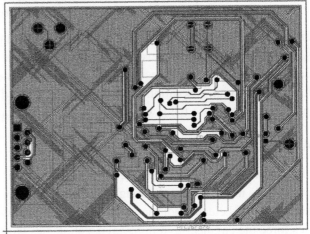

Figure 11.16: Bottom layer of an auto-routed layout of circuit diagram shown in Figure 11.17 and Figure 11.18.

With this layout, there are several areas of concern, but the most troubling issue is the grounding strategy. If you follow the ground traces on the top layer, the traces connect every device on that layer. A second ground connection for every device uses the bottom layer with vias at the far right-hand side of the board. The immediate red flag that one should see when examining this layout strategy would be the existence of several ground loops. Additionally, horizontal signal lines interrupt the ground return paths on the bottom side. The saving grace with this grounding scheme is that the analog devices (12-bit A/D converter and 2.5 V voltage reference) are at the far right hand side of the board. This placement ensures that digital ground signals do not pass under these analog chips.

Layout/Grounding (Precision, High Speed and Digital)

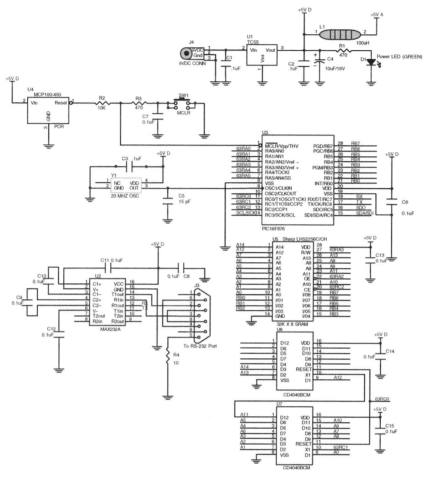

Figure 11.17: Digital section of circuit diagram for layouts in Figures 11.15, 11.16, 11.19 and 11.20. This is the circuit diagram from Microchip's MXDEV™ board, evaluation board for the 10- and 12-bit ADCs (MCP300X and MCP320x).

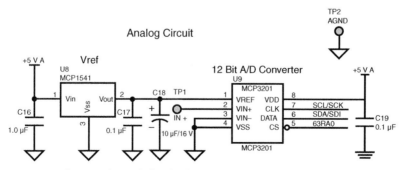

Figure 11.18: Analog section of circuit diagram for layouts in Figures 11.15, 11.16, 11.19, and 11.20. This is the circuit diagram from Microchip's MXDEV board, evaluation board for the 10- and 12-bit ADCs (MCP300X and MCP320x).

Chapter 11

Figure 11.19 and Figure 11.20 has the manual layouts of the circuits in Figure 11.17 and Figure 11.18. For the layout of this mixed-signal circuit, the devices were manually placed on the board with careful thought to separating the digital and analog devices. With this manual layout, a few general guidelines are followed to ensure positive results. These guidelines are:

1. Use the ground plane as a current return path as much as possible.
2. Separate the analog ground plane from the digital ground plane with a break.
3. If interruptions from signal traces are required on the ground-plane side, make them vertical to reduce the interference with the ground-current, return paths.
4. Place analog circuitry at the far end of the board and digital circuitry closest to the power connects. This reduces the effects of δi/δt from digital switching.

Note that with both of these two layer boards there is a ground plane on the bottom. This is only done so that an engineer working on the board can quickly see the layout when trouble shooting. You will typically find this strategy in manufacturer's demo and evaluation boards.

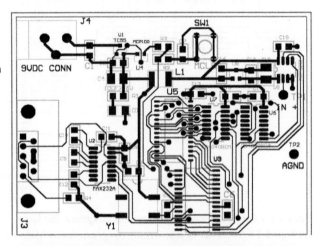

Figure 11.19: Top layer of a manual routed layout of circuit diagram shown in Figure 11.17 and Figure 11.18.

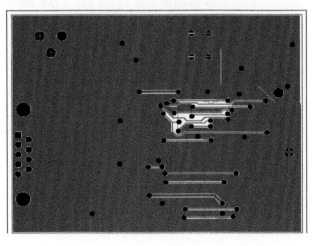

Figure 11.20: Bottom layer of a manual routed layout of the circuit diagram shown in Figure 11.17 and Figure 11.18.

Layout/Grounding (Precision, High Speed and Digital)

But more typically, the ground plane is on the top of board, thereby reducing electromagnetic interference (EMI).

At every layout-related presentation that I give in a seminar setting, the question always asked in one form or another is, "What if management tells me I can't have two layers or a ground plane, and I still need to reduce noise in the circuit? How do I design my circuit to work around the need for a ground plane?" Typically, I instruct the person asking the question to inform their management that a ground plane is simply required if they want reliable circuit performance. The primary reason for using ground planes is lower ground impedance. They also provide a degree of EMI reduction.

Current Return Paths With or Without a Ground Plane

The fundamental issues that should be considered when dealing with current return paths are:

1. If traces are used, they should be as wide as possible.

   In the event that you are considering using traces for your ground connects on your PCB, they should be as wide as possible. This is a good rule of thumb, but also understand that the thinnest width in your ground trace will be the effective width of the trace from that point to the end (where the "end" is defined as the point furthest from the power connection).

2. You should avoid ground loops.

3. If no ground plane is available, you should use a star connection strategy.

   Figure 11.21 shows a graphical example of a star connection strategy.

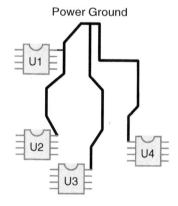

Figure 11.21: If a ground plane is not feasible, you should handle current return paths with a "star" layout strategy.

With this type of approach, the ground currents return to the power connection independently. You will note that in Figure 11.21 not all of the devices have their own return path. With U1 and U2, the return path is shared. This can be done if you use guidelines #4 and #5, following.

Chapter 11

4. Digital currents should not pass across analog devices.

   During switching, digital currents in the return path are fairly large, but only briefly. This phenomenon occurs due to the effective inductance and resistance of the ground. With the inductance portion of the ground plane or trace, the governing formula is $V = L\delta i/\delta t$, where V is the resulting voltage, L is the inductance of the ground plane or trace, δi is the change in current from the digital device and δt is the time span considered for the event. To calculate the effects of the resistance portion of the ground plane, changes in the voltage simply change because of $V = RI$. Again, V is the resulting voltage, R is the ground plane or trace resistance and I is the current change caused by the digital device. These changes in the voltage of the ground plane or trace across the analog device will change the relationship between ground and the signal in the signal chain.

5. High-speed current should not pass across lower speed devices

   Ground-return signals of high-speed circuits have a similar effect on changes to the ground plane. Again the more important formulas that determine the effects of this interference are $V = L\delta i/\delta t$ for the ground plane or trace inductance and $V = RI$ for the ground plane or trace resistance. And as with digital currents, high-speed circuits that have ground activity on the ground plane or that have a trace across the analog device change the relationship between ground and the signal in the signal chain.

6. Regardless of the technique used, you must design the ground return paths to have a minimum resistance and inductance.

7. If a ground plane is used, breaks in plane can improve or degrade circuit performance. Use with care.

But, if you are unable to win that battle with your management because of cost constraints, this book offers some suggestions. These suggestions are using star networks and current return paths, which if used properly, will give a little relief from the circuit noise.

Layout Tricks for a 12-Bit Sensing System

When I started writing this chapter I thought a "cookbook" approach would be appropriate when describing the implementation of a good 12-bit layout. My assumption behind this type of approach is that I would provide a reference design, which would make the layout implementation easy. But I struggled with this topic long enough to find that this notion was fairly unrealistic.

Because of the complexity of this problem, I am going to provide basic guidelines ending with a review of issues to be aware of while implementing your layout design. Throughout this discussion I will offer examples of good and bad layout implementations. I am doing this in the spirit of discussing concepts and not with the intent of recommending one layout as the only one to use.

Layout/Grounding (Precision, High Speed and Digital)

The application circuit that I'm going to use is a load-cell circuit that accurately measures the weight applied to the sensor, then displays the results on a LCD-display screen. Figure 11.22 shows the circuit diagram for this system. You have seen this circuit before in Chapters 3 and 10. This circuit was introduced in Chapter 3. In Chapter 10, noise reduction techniques were explored with this same circuit.

You can purchase the load cell that I used from Omega (LCL-816G). My sensor model for the LCL-816G is a four element resistive bridge that requires voltage excitation. With a 5 V excitation voltage applied to the high side of the sensor, the full-scale output swing is a ±10 mV differential-signal with a 32 ounce maximum excitation. A two-op amp instrumentation amplifier gains this small differential signal. I chose a 12-bit converter to match the required precision of this circuit. Once the converter digitizes the voltage presented at its input, the microcontroller receives the digital code by using the converter's SPI™ port. The microcontroller then uses a look-up table to convert the digital signal from the ADC into weight. Linearization and calibration activities can be implemented with controller code at this point if need be. Once this is done, the results are sent to the LCD display. As a final step, I wrote the firmware for the controller. Now the design is ready to go to board layout.

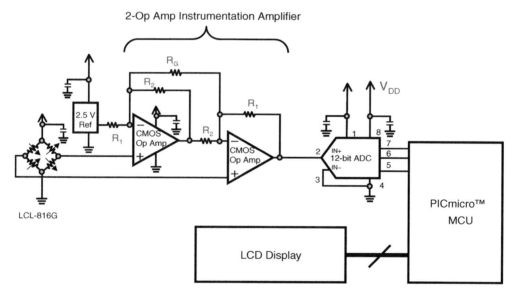

Figure 11.22: A two-op amp instrumentation amplifier, filtered and digitized with a 12-bit A/D converter, gains the signal at the output of the load-cell sensor. The result of each conversion is sent to the LCD display.

Chapter 11

General Layout Guidelines – Device Placement

My first step is to place the devices on the board. This critical step is done effectively because I am keeping track of my noise-sensitive devices and noise-creator devices. There are two guidelines that I use to accomplish this task:

1. Separate the circuit devices into two categories: high speed (>40 MHz) and low speed. You should place the higher speed devices closer to the board connector/power supply.
2. Separate the above categories into three sub-categories: pure digital, pure analog, and mixed signal. With this delineation, you need to place the digital devices closer to the board connector/power supply.

General Layout Guidelines – Ground and Power Supply Strategy

Once I determine the general location of the devices, I was able to define my ground and power planes. My strategy of the implementation for these planes is a bit tricky.

First of all, it is dangerous for me not to use a ground plane in a PCB implementation. This is true particularly in analog and/or mixed-signal designs. One issue is that ground noise problems are more difficult to deal with than power-supply, noise problems because analog signals are referenced to ground. For instance, in the circuit shown in Figure 11.22, the A/D converter's inverting input pin (MCP3201, Microchip) is connected to ground. Secondly, the ground plane also serves as a shield against emitted noise. Both of these problems are easy to resolve with a ground plane and nearly impossible to overcome if there is no ground plane.

However, with my small design, I assume that I won't need a ground plane. Figure 11.23 shows a ground plane-less, layout implementation of the circuit in Figure 11.22.

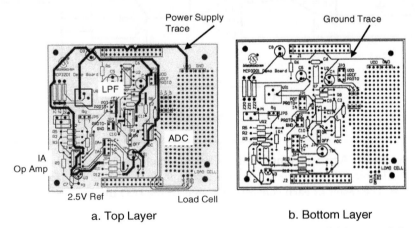

Figure 11.23: This is the layout of the top (a) and bottom (b) layers of the circuit in Figure 11.22. Note that this layout does not have a ground or power plane. Note that the power traces are considerably wider than the signal traces in order to reduce power supply trace inductance.

Layout/Grounding (Precision, High Speed and Digital)

Does my "no ground plane is required" theory play out? The proof is in the pudding, or data. In Figure 11.24, 4096 samples were taken from the A/D converter and logged. There was no excitation on the sensor when this data was taken. With this circuit layout, the controller is dedicated to interfacing with the converter and sending the converter's results to the LCD display.

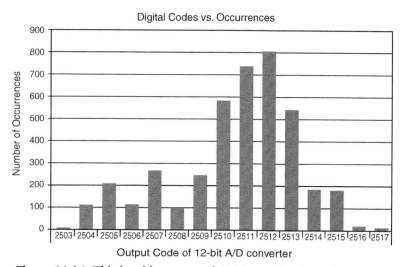

Figure 11.24: **This is a histogram of 4096 samples from the output of the A/D converter. The PCB does not have a ground or power plane as shown in the PCB layout in Figure 11.25. The code of the noise from the circuit is 15 codes wide (Figure 11.23).**

Figure 11.25 shows the same device layout shown in Figure 11.23, but a ground plane on the bottom layer is added. The ground plane (Figure 11.25b) has a few breaks due to signal. These breaks should be kept to a minimum. Current return paths should not be "pinched" as a

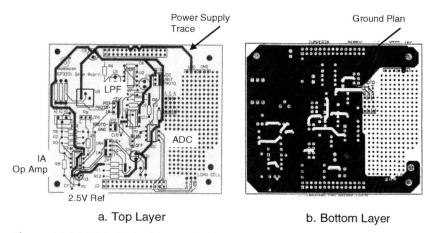

a. Top Layer   b. Bottom Layer

Figure 11.25: **This is the layout of the top and bottom layers of the circuit in Figure 11.22. Note that this layout *does* have a ground plane.**

285

consequence of these traces restricting the easy flow of current from the device to the power connector. Figure 11.26 shows the histogram for the A/D converter output. Compared to Figure 11.24, the output codes are much tighter. The same active devices were used for both tests. The passive devices were different causing a slight offset difference.

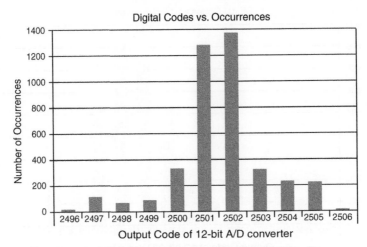

Figure 11.26: This is a histogram of 4096 samples from the output of the A/D converter on the PCB that has a ground plane as shown in the PCB layout in Figure 11.25. The code width of the noise is now 11 codes wide.

It is clear from my data that a ground plane does have an effect on the circuit noise. When my circuit did not have a ground plane, the width of the noise was ~15 codes. When I added a ground plane, I improved the performance by almost 1.5× or 15/11. You might want to know that my test set up was in the lab where EMI interference is relatively low.

The op amp and absence of an anti-aliasing filter causes of the noise shown Figure 11.26. If my circuit has a *minimum* amount of digital circuitry on board, a single ground plane and a single power plane may be appropriate. The board designer defines my qualifier *minimum*. The danger of connecting the digital and analog ground planes together is that my analog circuits can pick-up the noise on the supply pins and couple it into the signal path. In either case, I should connect my analog and digital grounds and power supplies together at one or more points in the circuit. This ensures that my power supply, input, and output ratings of all of the devices are not violated.

The inclusion of a power plane in a 12-bit system is not as critical as the required ground plane. Although a power plane can solve many problems, making the power traces two or three times wider than other traces on the board and by using bypass capacitors effectively can reduce power noise.

Signal Traces

My signal traces on the board (both digital and analog) should be as short as possible. This basic guideline will minimize the opportunities for extraneous signals to couple into the signal path. One area to be particularly cautious of is with the input terminals of analog devices. These terminals normally have a higher impedance than the output or power supply pins. As an example, the voltage reference input pin to the A/D converter is most sensitive while a conversion is occurring. With the type of 12-bit converter I have in Figure 11.22, my input terminals (IN+ and IN–) are also sensitive to injected noise. Another potential for noise injection into my signal path is the input terminals of an operational amplifier. These terminals have typically $10^9 \, \Omega$ to $10^{13} \, \Omega$ input impedance.

My high impedance input terminals are sensitive to injected currents. This can occur if the trace from a high impedance input is next to a trace that has fast changing voltages, such as a digital or clock signal. When a high impedance trace is in close proximity to a trace with these types of voltage changes, charge is capacitively coupled into the high impedance trace as mentioned earlier in the chapter.

Did I Say Bypass and Use an Anti-Aliasing Filter?

Although this chapter is about layout practices, I thought it would be a good idea to cover some of the basics in circuit design. A good rule concerning bypass capacitors is to always include them in the circuit. If they are not included, the power supply noise may very well eliminate any chance for 12-bit precision.

Bypass Capacitors

Bypass capacitors belong in two locations on the board: one at the power supply (10 µF to 100 µF or both) and one for every active device (digital and analog). The value of the bypass capacitor of the device is dependent on the device in question. If the bandwidth of the device is less than or equal to ~1 MHz, a 1 µF will reduce injected noise dramatically. If the bandwidth of the device is above ~10 MHz, a 0.1 µF capacitor is probably appropriate. In between these two frequencies, you could use both or either one. Refer to the manufacturer's guidelines for specifics.

Every active device on the board requires a bypass capacitor. It must be placed as close as possible to the power supply pin of the device as shown in Figure 11.25. If you use two bypass capacitors for one device, the smaller one should be closest to the device pin. Finally, the lead length of the bypass capacitor should be as short as possible.

Anti-Aliasing Filters

You will note that the circuit in Figure 11.22 does not have an anti-aliasing filter. As the data shows, this oversight has caused noise problems in the circuit. When this board has a 2nd order, 10 Hz, anti-aliasing filter inserted between the output of the instrumentation amplifier and the input of the A/D converter, the conversion response improves dramatically. Figure 11.27 shows the resulting data.

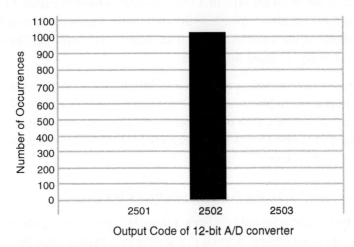

Figure 11.27: This diagram shows the conversion results of the circuit in Figure 11.22 plus a 2nd order, anti-aliasing filter. Additionally, the board layout includes a ground plane.

Analog filtering can remove noise superimposed on the analog signal before it reaches the A/D converter. In particular, this includes extraneous noise peaks. Analog-to-digital converters will convert the signal that is present on its input. This signal could include the sensor voltage signal or noise. The anti-aliasing filter removes the higher frequency noise from the conversion process.

PCB Design Checklist

Good layout techniques are not difficult to master as long as you follow a few guidelines:

1. Check device placement versus connectors. Make sure that high-speed devices and digital devices are closest to the connector.
2. Always have at least one ground plane in the circuit.
3. Make power traces wider than other traces on the board.
4. Review current return paths and look for possible noise sources on ground connects. Determining the current density at all points of the ground plane and the amount of possible noise present does this.

5. Bypass all devices properly. Place the capacitors as close to the power pins of the device as possible.
6. Keep all traces as short as possible.
7. Follow all high impedance traces looking for possible capacitive coupling problems from trace to trace.
8. Make sure you properly filter your signals in a mixed-signal circuit.

Analog layout and digital layout techniques differ slightly, but not completely. When it comes to the parasitic components embedded in the PCB, the analog circuits tend to show more sensitivity, but digital circuits are not completely immune. You should treat the device that straddles these two domains, such as the A/D converter, as an analog device. Two-layer boards do present some challenges, but careful, manual layouts can usually work around these problems. You can fix a poor layout if you are willing to go back to the drawing board.

When the analog and digital domains meet, careful layout is critical if a designer intends to have a successful final PCB implementation. Layout strategies usually are presented as rules of thumb because it is difficult to test the success of your final product in a lab environment. So, generally speaking, although there are some similarities in layout strategies between the digital and analog domain, the differences should be recognized and worked with.

Solving signal integrity problems can take a great deal of time, particularly if you don't have the tools to tackle the tough issues. The three best analysis tools to have in your arsenal are the frequency analysis (Fast Fourier Transform or FFT), time analysis (scope photo), and DC analysis (histogram) tools. We used all of these tools to identify the power supply noise, external clock noise, and overdriven amplifier distortion.

Chapter 11 References

MXDEV is a trademark of Microchip Technology Inc. in the USA and other countries.

SPI™ port is a trademark of Motorola. The Microchip name and logo, PIC, PICmicro, microID and KEELOQ are registered trademarks of Microchip Technology Inc. in the USA and other countries. All other trademarks are the property of their respective owners.

Noise Reduction Techniques in Electronic Systems, 2nd ed., Henry W. Ott, Wiley, 1998.

Noise and Other Interfering Signals, Ralph Morrison, John Wiley & Sons, 1992.

"Circuit Layout Techniques and Tips: 6 Part," Baker, Bonnie C., First Published in *analogZone* (2002, 2003) and reproduced with permission.

CHAPTER 12

The Trouble With Troubleshooting Your Mixed-Signal Designs Without the Right Tools

CHAPTER

12

The Trouble With Troubleshooting
Your Mixed-Signal Designs
Without the Right Tools

CHAPTER 12

The Trouble With Troubleshooting Your Mixed-Signal Designs Without the Right Tools

When you're trying to solve a signal integrity problem, the best of all worlds is to have more than one tool to examine the behavior of your system. If there is an analog-to-digital converter (ADC) in the signal path, there are three fundamental perspectives that you can take when assessing the circuit's performance. All three of these angles evaluate the conversion process, as well as its interaction with the layout and other portions of the circuit. These three areas encompass the use of frequency analysis (Fast Fourier Transform, FFT), time analysis, and DC analysis techniques. This chapter will explore the use of these tools to identify the source of problems as they relate to the layout implementation of circuits. We will explore how to decide what to look for, where to look, how to verify problems through testing, and how to solve the problems that you find.

The Basic Tools for Your Troubleshooting Arsenal

In your laboratory, you will need some of the basic tools. As a starting point, I would suggest that you equip yourself with a soldering iron, a multimeter, and a desktop oscilloscope. You might comment that this not enough to do the job. You may be right, but this type of equipment will get you a long way to finding your circuit problems. There is more, but for now let's talk about these three things.

You probably figured out that you needed a soldering iron to build your circuit. But, more than that, you will use it to verify that all of your junctions are properly soldered. I have run into this problem more than you would think. For example, a circuit doesn't work at all or generates signals that are very unexpected or unusual. If you are good with the soldering iron, you won't need to return to this task and touch things up. God bless the technicians that you have on staff. They will probably get it right long before you will. But, if (on occasion) you (or they) miss joints or you don't make the solder connection the first time, this tool will save you an untold amount of time.

The multimeter is the second tool that I recommend. You can purchase a handheld meter that will perform admirably. Be careful, though. Don't scrimp and save on money by getting the low-end equipment. You will find that your low-end meter will take measurements but they will be wrong. This is an unfortunate situation for those who trust their equipment implicitly.

Chapter 12

For instance, I was working in a lab where we were trying to measure a voltage across a resistor divider. One resistor was external and one was an internal resistor on an amplifier chip. We consistently measured too low of a voltage value. We were also using an inexpensive voltmeter. One of my co-workers exclaimed, "It looks like the chip vendor didn't know what he was doing. His specification in the data sheet for the value of the resistor is wrong."

We discovered, after reading the equipment manual, the voltmeter's output resistance was only 100 kΩ. That was a new one for me. I was accustomed to using voltmeters with at least 10 MΩ if not higher. The resistor network values that we were trying to measure were in the tens of kilo-ohms. Our test equipment (the voltmeter) was changing the circuit by adding another resistor to the system. So the moral to this story (for me) was don't scrimp on your equipment. Secondly, know the specifications of your equipment so you can understand and verify that you are getting reliable, instead of misleading results.

When you apply power to the board, use the multimeter to verify DC voltages across your board. Connect the negative probe to the ground on the board. Check that the power voltages actually have the correct voltages by touching your probe to the power traces. Touch your probe on the supply pins of the devices on your board, comparing these pins to ground. Notice that I didn't say touch the probe to the power traces, but to the actual device pins. Once you verify that the devices have the correct power, go through the board and test the other DC voltages, such as references or voltage dividers, etc.

The third piece of equipment that I recommend is a desktop oscilloscope. I am particularly fond of digital scopes. As a side comment, you sometimes will find that the analog scopes will give you a better picture, minus aliasing problems, but the digital scope gives you features that are very handy. You will also use the oscilloscope to measure dynamic analog signals on your board. These dynamic signals may be your input signal, clocks, etc.

For instance, have you ever tried to look at a serial signal coming out of an A/D converter? You might say that it is pretty easy to do, but have you ever tried to deduce the value of each code? With a digital scope, you call look at every bit and average each bit with several conversions. My bench scope will average up to 128 times. That is more than enough for me to determine if one of the many codes coming out of the converter has a major error. This is very convenient if you are aware that the A/D converter output code is changing on the board, but can't identify the place where the error has occurred. For instance, has it occurred with the converter? or has it occurred coming out of your memory chip? Or some place else?

My final recommendation for lab equipment is the vendor's demo or evaluation board. The vendor boards have been carefully laided-out so that their device is at its best. If you are lucky, the vendor will translate the output of the device they are trying to promote, onto your computer screen. You will find tools such as histograms, scope plots, or FFT results, to name a few. These data results are usually digitally taken at the output of the vendor's device, but you can breadboard (or dead-bug wiring) any circuit you want onto their board.

The Trouble With Troubleshooting Your Mixed-Signal Designs Without the Right Tools

These boards will hopefully give you a tool that will display multiple (DC input) samples from your system over time in a histogram plot. In this manner, you will take a DC measurement of your system in a different way. The multimeter can only give you a sample of an averaged voltage. A histogram of several samples will enhance your view of what is happening in your circuit. This enhancement of the DC measurement will give you information about your system over time. The over-time-data that is of interest are drifts or changes over temperature and time.

These tools will get you through any fundamental evaluations that you undertake. You may need to go further, at which point you need to find the right tool to solve your particular problem. As we go through the examples in this chapter, we are going to use some of these tools. Remember that these recommendations are only the basics. If you feel you need higher-priced, more-capable, equipment, by all means, get them.

You ask, "Does my Circuit A/D Converter Work?"

I am using the circuit in Figure 12.1 for the following discussions. You have seen this circuit before in Chapter 3 (Figure 3.8), Chapter 10 (Figure 10.5), and Chapter 11 (Figure 11.22). In Chapter 3, we talked about how to use this circuit for pressure or load sensing applications. This circuit efficiently converts a differential analog signal into a single ended solution. It is not easy to do this differentiation task with digital circuits. This is because you must simultaneously sample the differential signal. If you decide to sample each leg of the pressure sensor independently and then subtract you two samples, you need to sample at a very high speed to reject common-mode noise. Rejection of common-mode noise as low as 50 Hz or 60 Hz requires a very, high-speed conversion time (~6.2 Msps for a 60 Hz rejection to 0.25 LSB accuracy). You can use your lab equipment and your vendor's boards to test this theory.

In Chapter 10, we talked about noise reduction and this same circuit was used to demonstrate the advantages of analog filters. In the discussion in Chapter 10, we decided (theoretically) that the placing of analog, low-pass filters before the A/D converter removes higher frequency signals. These higher frequencies could be aliased into the output, digital word. The results at the output of the A/D converter over time would look noisy in the lower bits. With the equipment mentioned could you or prove this in your lab? In Chapter 11, we examined layout options with this circuit. We found out the advantages of using a solid ground plane and separating digital circuitry from analog.

Let's revisit this circuit. This discussion is where we will find out if all the theories discussed in Chapter 3 and Chapter 10 are true.

For testing purposes, two inputs to the instrumentation amplifier (V_{IN+} and V_{IN-}) are equal to 2.5 V DC. The power supply voltage and reference to the A/D converter is equal to 5 V. A 5 V-voltage reference applied to the A/D converter creates an LSB size of the 12-bit converter equal to 1.22 mV. Given these conditions I expect the digital output code from the A/D

Chapter 12

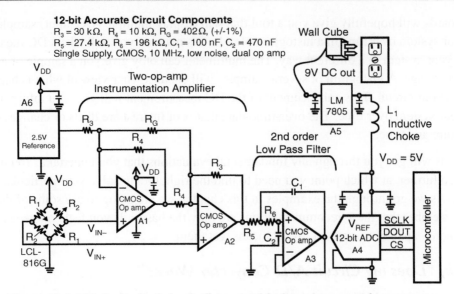

Figure 12.1: This is the final circuit diagram with the necessary enhancements added. The instrumentation amplifier (A_1 and A_2) gains the voltage at the output of the SCX015 pressure sensor. Following the instrumentation-amplifier. a low-pass filter (A_3) is inserted to eliminate aliased noise from the 12-bit ADC conversion. All of the active devices have by-pass capacitors and L_1 filters the power signal from the "wall cube." This should attenuate high frequency noise.

converter to equal (2.5 V/1.22 mV) or 2049. The purpose of the following tests is to verify that the 12-bit conversion is noise free. This means that the converter will produce one code for every sample. If there isn't amplifier offset errors or with the 2.5 V reference, this one code will be equal to 2049.

Figure 12.2 illustrates the layout for the final circuit in Figure 12.1.

This circuit had many enhancements in an attempt to make it convert with 12-bit repeatability. Each step of the way, tests were preformed on the circuit in the lab to verify that changes guided by theory was correct. For instance, I thought that the initial resistors around the instrumentation amplifier were too high. This hunch was verified in SPICE simulations (Figure 12.3) and then on the bench.

In Chapter 10, we looked for noise sources using SPICE simulations in the instrumentation amplifier of this board. In simulation we verified that higher value resistors caused noise (R_3 = 300 kΩ, R_4 = 100 kΩ, R_G = 4.02 kΩ). This was not a surprise. However, the theories in Chapter 10 need to be verified on the bench. This verification was done with histogram plots. The lower-noise resistors around the instrumentation amplifier (R_3 = 30 kΩ, R_4 = 10 kΩ, R_G = 402 Ω) were compared to the higher value resistors. Histogram plots of this improved circuit demonstrated a small improvement in noise. The histogram plots showed a change in the standard deviation from 0.375 codes to 0.346 codes.

The Trouble With Troubleshooting Your Mixed-Signal Designs Without the Right Tools

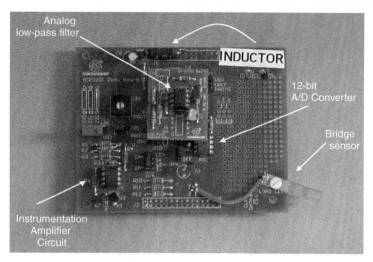

Figure 12.2: This figure shows the final layout of the circuit in Figure 12.1. Using lab tests you can validate and verify changes in the design. On this board, this circuit starts the signal starts from the load-cell and finishes at the A/D converter output. Data is transmitted off this board, processed by a microcontroller, and sent to the computer through an RS-232 port (not shown here). You will find an analog low-pass filter, by-pass capacitors, and power-supply choke (L_1) on this board. All three of these enhancements plus the reduction of the noisy resistors around the instrumentation-amplifier reduces the noise enough so that this circuit becomes a TRUE 12-bit circuit. I verified these enhancements on the bench in my lab, using the vendor's (Microchip) demo-board and interface software.

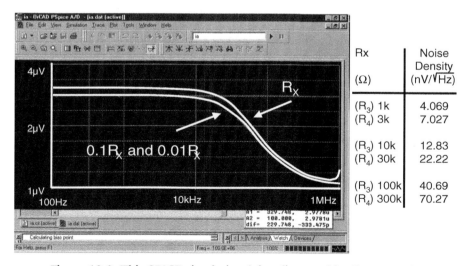

Rx (Ω)	Noise Density (nV/$\sqrt{Hz}$)
(R_3) 1k	4.069
(R_4) 3k	7.027
(R_3) 10k	12.83
(R_4) 30k	22.22
(R_3) 100k	40.69
(R_4) 300k	70.27

Figure 12.3: This SPICE simulation (also discussed in Chapter 10) suggests that the resistors in the circuit in Figure 12.1 cause noise. The resistors on the board were changed. Histogram test results verified this small improvement in the repeatability of this circuit.

Chapter 12

Power Supply Noise

A common source of interference in circuit applications is from the power supply. The power supply pins of the active devices receive this noisy signal from the power-supply. In the lab, this was a tough one to track down and I will show you why. Figure 12.4 shows the starting point for this circuit.

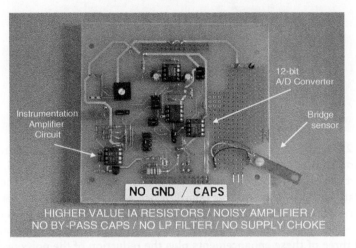

Figure 12.4: There were numerous problems with the first PCB implementation of the circuit in Figure 12.1. In this first board revision, the resistors around the instrumentation amplifier were higher value resistors (R3 = 300 kΩ, R4 = 100 kΩ, RG = 4.02 kΩ). Noisy amplifiers were used for the two op amps in the instrumentation amplifier. The by-pass capacitors were missing and there was no low-pass filter in the circuit. Finally, the power supply was not filtered with a choke inductor, L_1 (per Figure 12.1).

A histogram of the output of the A/D converter in Figure 12.1 on the board if Figure 12.2 (minus the inductor (L_1)) is given in Figure 12.5. In Figure 12.5, the sample speed for the ADC was 40 ksps for the 4096 samples.

If you look at the data in Figure 12.5 you would assume that there was "random noise" is in the system. But, is this true? When we looked at this data in the lab many of the engineers were very eager to say that the converter was the problem. And I would say that this is a fine assumption; but prove it.

The next stop in this analysis was the FFT plot, which was a view of the data in the frequency domain. This would tell us if we were actually looking at random noise. The results from the FFT plot (Figure 12.6) were a surprise to us all!

This "random" noise that we first observed in the histogram was showing anything but a random pattern in the FFT (see Figure 12.6). The noise appears to have a fundamental frequency or 1982.6 Hz with a complex overtone series.

The Trouble With Troubleshooting Your Mixed-Signal Designs Without the Right Tools

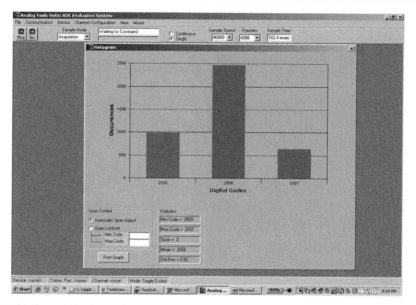

Figure 12.5: This is a collection of 4096 data outputs from the 12-bit A/D converter in Figure 12.1 The data rate of these 4096 data points is 40 ksps. From this histogram one could assume that the A/D converter or the analog signal path before the converter was noisy.

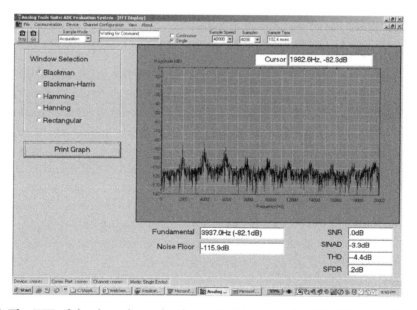

Figure 12.6: The FFT of the data shown in Figure 12.5 is a surprise. The histogram in Figure 12.5 uses the same data, but in a histogram representation instead of the FFT above. We expected random noise, which would look like a horizontal band across this graph. This data is actually showing that there is a signal with a fundamental frequency of approximately 1982.6 Hz. This data suggests that the once assumed "random" noise is anything but that.

Chapter 12

The last step in this evaluation was to look at the data in the time domain (scope plot) as opposed to the frequency domain (FFT). Figure 12.7 shows those results.

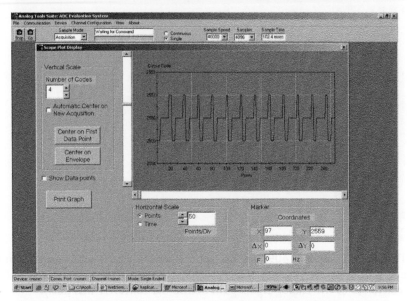

Figure 12.7: The time domain representation of this data from the 12-bit A/D converter produces an interesting periodic signal. This signal was traced back to the power supply. Each graph in Figure 12.5, 12.6, and 12.7 uses the same data.

Further investigations into the circuit show that the source of the noise seen in the scope and FFT plots come from the switching power supply (the wall cube). You ask, how did I know this? I really didn't know for sure. After looking at all of the components on the board and the PCB traces in the layout (see Chapter 11), the power-supply signal was the most likely suspect. An inductive choke (L_1) was added to the circuit along with bypass capacitors. One 10 µF was positioned at the power supply and 0.1 µF capacitors were placed as close to the supply pins of the active elements as possible. Now the generation of a new time plot seems to produce a solid DC output and this proven with the histogram results is shown in Figure 12.8. We couldn't look at the FFT plot because we were producing only one code at the output of the A/D converter. The data shows that these changes eliminated the noise source from the signal path of the circuit.

There were ten steps in this troubleshooting process. We first verified the DC voltages across the board. Every thing looked good. Then we looked at the multiple DC samples at the output of the A/D converter (Figure 12.5) in a histogram format. This DC data was collected using vendor's software and the A/D converter's evaluation board. We threw around theories at this point, as to why the signal was noisy, but nothing was completely substantiated.

300

The Trouble With Troubleshooting Your Mixed-Signal Designs Without the Right Tools

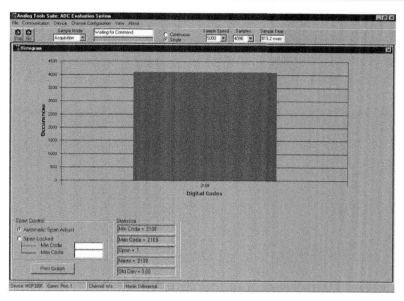

Figure 12.8: Once the power supply noise has been sufficiently reduced, the output code of the 12-bit A/D converter is consistently one code, 2108.

The next step was to look at an FFT plot of raw data from the histogram plot. We could have very well gone to the scope plot (and maybe saved time), but the FFT plot did show some disturbing results. There was a signal with frequency content in our data, not the noisy, random noise that we thought at first (Figure 12.6). At this point, we had enough data to get considerably closer to finding the noise generator in the circuit. But call me obsessive, I wanted to see this conclusion in another format.

We then looked at the scope plot of the same data (Figure 12.7). With three different representations of the same data saying the same thing, we were half way to a solution to this noise problem. Now we started to look in earnest for the culprit device(s). We decided the power supply wall cube was the problem. I then filtered the power supply with L_1 and bypass capacitors at all of the active device's power pins to eliminate noise from the power supply and eureka. Problem solved (Figure 12.8).

This process is tedious but complete. Use your tools to investigate your theories. Don't use your tools to collect data for data's sake. Were we done with this circuit? No. We were ready to proceed to the next level; temperature tests.

Improper Use of Amplifiers

Returning to the circuit shown in Figure 12.1, a 1 kHz AC signal is injected at the positive input (V_{IN+}) to the instrumentation amplifier. This signal would not be characteristic of the load-cell sensing circuit shown in Figure 12.1, however, this example is used to illustrate the

Chapter 12

influence of the front-end analog circuits in the analog signal path. Figure 12.9 shows the performance of this circuit with the above conditions in the FFT plot. You should notice that the noise floor is significantly higher as compared to any other FFT plot shown thus far. Additionally, the spurs have a multi-frequency characteristic that could again be difficult to track down.

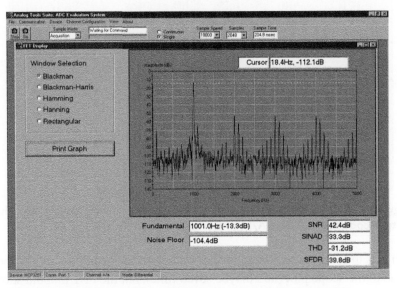

Figure 12.9: This FFT shows a distortion in the circuit that we haven't seen before. The only change to the circuit was injecting a signal into the input of the instrumentation amplifier. Would you guess that the external signal has distortion built-in, the amplifier is distorting the signal, or the A/D converter is causing this distortion?

This FFT plot does not have the same characteristics as before (Figure 12.6). The distortion in the FFT plot of Figure 12.9 has a fundamental frequency of 1 kHz. This was verified with the oscilloscope by testing the input signal at V_{IN+}. These results were at the same frequency as the input signal. A good place to look when troubleshooting this kind of issue is the signal chain and the behavior of each element in that chain. The histogram of this output data from the A/D converter will not tell you much. You can track down this distortion problem with a time-domain or scope plot constructed from the output codes of the A/D converter. Figure 12.10 shows the scope plot of this output data, along with an insert.

At first glance, the sine wave that is created by the A/D converter output code looks fairly good (see Figure 12.10). But, as you look at the lower portion of the sine wave closer, you will notice that the signal is sometimes clipped (see insert in Figure 12.9). This clipping occurs in the operational amplifier. The solution to this problem is to level shift the input signal to a slightly higher value. You could also achieve this by adjusting the reference voltage to the instrumentation amplifier, A_6.

The Trouble With Troubleshooting Your Mixed-Signal Designs Without the Right Tools

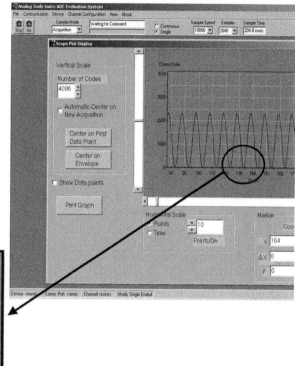
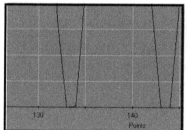

Figure 12.10: The time domain representation of the raw data from the FFT plot in Figure 12.8 shows that one of the amplifiers in the circuit is clipping the input signal on the negative rail. The sample rate of this test is 10 ksps. This graph has 2048 samples.

Don't Miss the Details

You can get analog design experience through many channels. One channel that I am particularly found of is hands-on. Sometimes I kid around by telling my audiences in seminars that an engineer has done some hands-on work if they have scar tissue on their fingertips because they just didn't believe the circuit they are working or is getting that hot. This comes from looking at a problem circuit, asking what has gone wrong and then touching every chip.

Another tell-tale-sign is when a designer has dead bug circuits in his workspace. This type of circuit is built by turning the device on its back, with pins sticking straight up and soldering all connections to the device in the air. Using this technique is an attempt to eliminate parasitic capacitance.

Yet, another indication that an individual spends time in lab is facial scar tissue that looks like little razor cuts. This may indicate a problem with shaving, but more than likely the designer was looking too closely at a circuit when a capacitor or chip "blew up." With these kinds of

Chapter 12

experiences, knowledge comes. The first thing you will quickly learn is to use eye protection in the lab. Following are a few additional tips that I have learned.

Double-check your power supplies before you plug the chip in. You need to double-check them with a voltmeter. I have seen a lot of engineers assume that they have the right value voltage on their power supply pins only to find out that they are wrong. I also see engineers blindly trusting the LCD read out of their power supplies. They trust these read-outs so much they use them in their calculations while they are looking for problems and solutions.

You can damage active devices if the positive supply is too high at any time with respect to the negative supply. In contrast, a low supply won't bias the amplifier's or A/D converter's internal transistors. A simple check of the difference between the supply-voltages at the pins of your device, can save a great deal of troubleshooting time. Just a note of caution, turn the supplies off before you remove or insert any active device into their sockets.

Double-check your grounding strategy, especially if there are digital circuits on the board. Low impedance grounds are imperative if you want a stable analog design. If the circuit has a lot of digital circuitry, consider separate ground and power planes. Ground noise is a real challenge to track down because it seems to be everywhere. Along with this discussion goes the unintentional creation of ground loops. Creating ground loops in your layout is one issue, but I have seen engineers create ground loops with the test equipment leads. On one occasion I saw an engineer create seven different returns to different grounds with the test equipment leads. It turned out he was measuring noise emissions that were picked up by the test leads hanging in the air.

Always bypass the amplifier power supply pins with capacitors. I know that I have mentioned this before in other chapters, but saying it more than once can't hurt. Place these capacitors as close to the amplifier pin as possible. I usually recommend a 1 µF or 0.1 µF capacitor for amplifiers that have a bandwidth up to tens of megahertz. Look for the existence and location of bypass capacitors while troubleshooting your circuits on the bench.

Breadboarding on white perf boards is a risky way of doing your circuit evaluation business. These boards can produce noise or oscillations because of the preponderance of capacitance and inductance hidden underneath the board. Since you should use short lead lengths to the inputs most of the analog devices you are using, the perf board will fail you. This problem is aggravated as your circuit operation goes up in frequency. Any circuit operating above 1 MHz is a problem. These higher frequencies are not reserved for high frequency analog devices. The digital devices have a preponderance of high frequency signals. You might think that your clock frequency is the highest frequency on your board, but the rise and fall times of your microcontroller gates create significantly higher frequencies on your board. There is a good chance that these problems won't be a problem with the PCB implementation of the circuit.

Analog devices are static sensitive! If they are damaged, they may fail immediately or exhibit a soft error (like offset voltage or input bias current changes) that will get worse over time.

Conclusion

If you don't know what the problem is, you can't fix it. Sometimes we see a problem suddenly "go away." You may hand-wave an explanation for the miraculous healing power of your circuit, such as the code changed or the layout changed or the season changed (lowering the humidity) and so forth and still not know what really happened. My recommendation is to verify and validate your theories from two or three different test angles. Solving signal integrity problems can take a great deal of time, particularly if you don't have the tools to tackle the tough issues. The three best analysis tools to have in your arsenal are the frequency analysis (FFT), time analysis (scope photo), and DC analysis (histogram) tools. We used all of these tools to identify the power supply noise, external clock noise, and overdriven amplifier distortion.

Chapter 12 References

Noise Reduction Techniques in Electronic Systems, Ott, Henry, John Wiley and Sons, 1988.

Analog-Digital Conversion Handbook, Sheingold, Daniel, Prentice Hall, 1986.

Mixed-Signal and DSP Design Techniques, Analog Devices, 2000 Seminar.

http://www.tek.com/Measurement/App_Notes/fft/.

http://www.bores.com/qedesign/tech/.

"Sources of Spurious Components in a DDS/DAC System," Crook, Cushing, April, 1998, RF Design, p. 28.

A Simple Approach to Digital Signal Processing, Marven, Craig, Ewers, Gillman, John Wiley, New York.

"Circuit Layout Techniques and Tips: 6 Part," Baker, Bonnie C., First published in *analog-Zone* (2002, 2003) and reproduced with permission.

CHAPTER 13

Combining Digital and Analog in the Same Engineer, and on the Same Board

CHAPTER 13

Combining Digital and Analog in the Same Engineer, and on the Same Board

Sometimes the digital engineer looks at an analog signal with a degree of discomfort. In analog, there are uncertainties that are not easy to explain—for example, a noisy output result from your analog system. Can you always explain it? Take heart. Sometimes even the good analog engineer doesn't know why.

I have also met digital engineers that scrutinize every conversion bit and every nano-volt. I know that this sounds anti-digital. Some digital engineers are just getting started. Others have a great understanding about what they are trying to do with their analog systems. For example, I visited a company in Indiana, USA. These digital engineers worked through a complete error analysis that most analog engineers have yet to aspire to. Digital engineers were solving this problem because the ADC was internal to the microcontroller. I spent a complete day with these engineers, only to find out they would not be using our parts. Go figure. But they knew what they were doing.

On the other hand, the analog engineer can look at digital translations with some degree of uneasiness. But there may easily be instances where digital and analog engineers think they know more than they really do. For instance, in the beginning of my career I was answering the engineering hot-line one day, for my (analog) company. I picked up a customer that was asking if my company had a 32-bit ADC. This question perplexed and fascinated me. Instead of saying, a flat-out "No" (which is bad business), I asked the engineer on the phone why he needed such a converter. He said that he had a 32-line bus that he needed to interface to the converter. You can guess the solution to the customer's problem. My first question was whether this customer was analog or digital centric.

I suggested that he find a less-expensive converter (i.e., 12-bit or 16-bit) and tie the unused bus lines to ground. I also said that I was not aware of any 32-bit ADC on the market, at least at that time. This is still true, to my knowledge.

The Signal Chain to the Real World

The point of this book is to learn how to interface your circuits to the real-world. Figure 13.1 shows a generic signal chain that describes a possible signal path. You've seen this figure in Chapter 1 (Figure 1.19), but by now we have a little better knowledge of what is involved in this diagram. In this figure, you will find the sensors at the far, left side of the diagram. By progressing to the right in this figure, the analog amplifier stage and anti-aliasing filter prepares the small sensor signal for the input of the ADC.

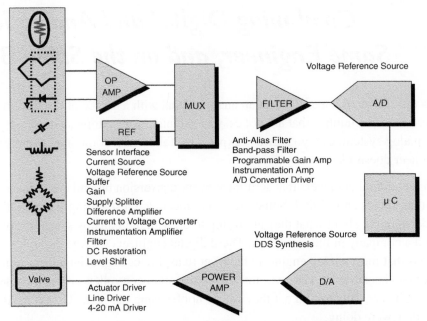

Figure 13.1: This diagram contains a classic sensor analog signal chain. There are a variety of sensors on the left side of this diagram. The amplifier stage conditions and the anti-aliasing circuit filters the sensor signal in preparation for the input of the ADC. The microcontroller device acquires that digital output code from the ADC.

There are a variety of implementations of this signal chain. This variety ranges from what is shown in Figure 13.1, to a simple R/C pair, which is connects the sensor to the microcontroller.

Tools of the Trade

This book starts out with an analog-to-digital converter discussion. I have accepted some criticism for starting with the converter, and I think that most books start with the operational amplifier. Besides the point that there is an amplifier in almost every analog device, this still doesn't make sense to me. I guess I am starting to see the "digital light." If you are digital centric, I think that you should start your analog learning with the device that is closest to the microcontroller or microprocessor. This seems more beneficial and practical to me. The ADC is easy enough to understand from the digital side. Once modeled with R/C circuits, the analog interface is also manageable.

So, the beginning of this book lists the most common converters that the digital designer uses when measuring real-world entities. These converters are the successive approximation register (SAR) ADC and the sigma-delta (Σ–Δ) ADC. With these converters, the key areas that you should concentrate on are the input driving devices, sampling time, noiseless input signals, voltage range, etc.

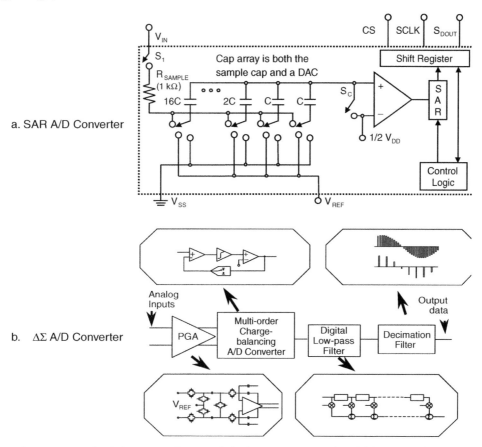

Figure 13.2: The (a) SAR analog-to-digital converter acquires high speed signals that only need minimal resolution. The (b) Σ–Δ analog-to-digital converter acquires a signal for higher resolution at slower speeds than the SAR converter.

The quandary that follows, the basic introduction of ADCs is interfacing to the real-world. The task is to capture real-world events, record, and use that information to control or affect things by using the microcontroller or processor. This can be straightforward or incredibly challenging. For instance, I had a customer (again on the engineering hot-line) that was trying to design a single system where he could measure the output of any sensor. I thought that this would be easy enough, but his problem was that he didn't know anything about electronics. I suggested he learn a few fundamentals, for example, Ohm's Law, which I immediately launched into.

Chapter 13

He exclaimed that he was not interested in that kind information. He explained there presently was no solution for this problem, because everyone was stuck in the same paradigm. He was going to break that trend by choosing to solve this problem outside the "box." In his estimation, electronic theory was the roadblock. At this point, both of us became frustrated. He felt that I could not answer his questions and I thought a slight adjustment in his perspective would go a long way.

One of the keys to working in analog is to use the appropriate low-pass filter before the ADC. This could be a simple R/C filter, or an amplifier active filter. At any rate, without exception, you place this filter before the ADC in the analog domain. If you try to use a digital filter exclusively to get rid of your unwanted noise in your signal chain, good luck. Don't neglect the analog performance to the last minute. You will find that the extra value that you will invest in analog devices will amount to a fraction of the added digital processing time and cost. Simply put, if you go from analog to digital you will need to filter the noise in your system. Remember that the electronics will not read your mind and know that you are only interested

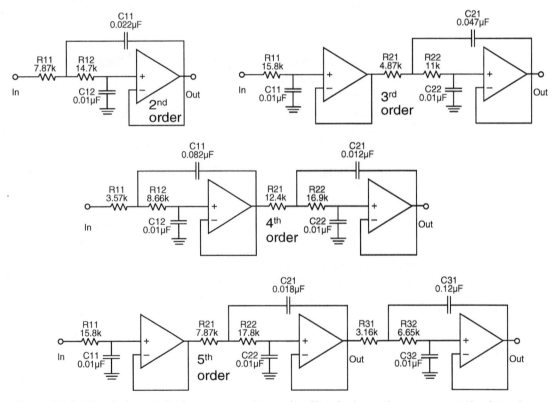

Figure 13.3: The design of the low-pass analog active filter is dependent on your selection of resistors and capacitors. You can design 1st order, 2nd order, 3rd order, 4th order, 5th order or higher. These are all Butterworth filters with a corner frequency or 1000 Hz and generated by a software filter program.

in signals between certain frequencies. The electronics will succumb to the physics of your circuit, and all passive and active devices have noise across a broad range of frequencies!

Along with the filter, you need to look for the amplifier. The amplifier is not only contained in almost every analog device, but also used extensively, externally. The fundamental circuits that you will use at one time or another, are the buffer, difference amplifier, summing amplifier and current-to-voltage amplifier.

But in this tour through the analog world, tools come to mind. SPICE programs, this valuable simulation tool, can enhance your analog development and circuit quality. You should use it with caution and like any design, question the results. SPICE programs do not relieve you of the task of thinking. It just helps you catch things you didn't think would make a difference in your circuit.

Now to even out your tools, you are not limited to digital side of your microcontroller or microprocessor. There are enough new peripherals in the micros that they are starting to "take-over" the analog domain. The only thing left behind is the high precision analog stuff. But is it still being left behind today in the product offerings? I think that I have seen high-resolution, $\Sigma-\Delta$ converters embedded in a few microcontrollers.

Noise is an interesting gremlin to grapple with. If you don't know where it originates you will be hard pressed to reduce it to a tolerable magnitude. So here is the task. Find the origin. Then reduce the noise by using new components for device noise, filtering for conducted noise, and layout tricks for radiated noise.

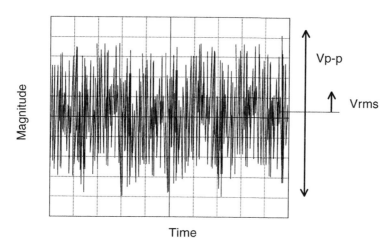

Figure 13.4: This is an oscilloscope picture of random noise. If your noise is gaussian, you can characterize and predict possible outcomes from your circuit by using statistics. The rms value of your sampled data is equivalent to one standard deviation.

Chapter 13

You can attribute circuit noise to many facets in your application circuit. It is connected to device selection, conducted noise on the board and radiated noise. Layout is a tool that you can use to battle the entrance of noise into your circuit. Ground planes, power planes come into play. Device location and relation to other devices is an issue. This is mixed into the other areas of your circuit. It is sometimes difficult to determine where the noise is coming from, so using good techniques from the beginning is a good start.

At any rate, try as you may, theory will only take you so far. If you don't validate the operation of pieces or all of your circuit, you will be disappointed. In my career, I have only known one engineer that I thought was lucky. He always landed on his feet, like a cat. Come to find out, he knew his stuff and had it proven in the lab. So much for counting on luck to get you through the tough times. I know, for certain, the things I accomplish are not a matter of luck, but hard work. Always check your circuits. Don't trust the theory and really don't trust your intuition without solid data. The histogram, scope, FFT, multimeter are all valuable tools for your personal bench.

Throwing the Digital In With the Analog

Imagine you are going to design an analog/digital circuit for your company. Figure 13.5 shows a block diagram of such a circuit.

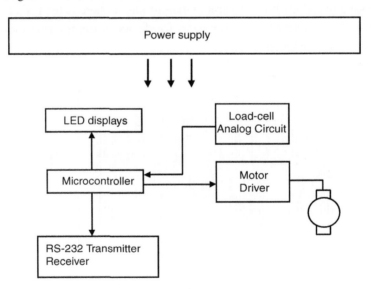

Figure 13.5: This block diagram models the circuit, along with noise sources, of the system in this book. The analog interface circuitry measures weight with a load-cell sensor. The interface then transmits those results to a microcontroller. The microcontroller sends the sensor results to an LED display and laptop computer. There is also circuitry to monitor the board temperature. In the event that the board temperature becomes too high, the microcontroller actuates the motor driver, which excites a fan.

Figure 13.6 illustrates the circuit diagram of this analog section.

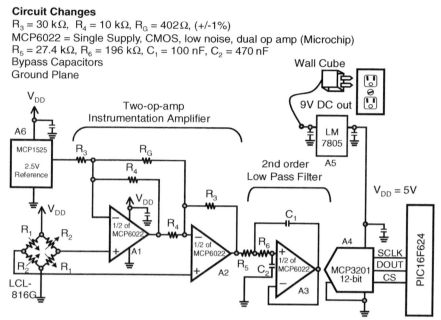

Figure 13.6: This is the analog portion of the block diagram drawing in Figure 13.5. The amplifiers and voltage reference tie into a microcontroller, PIC16F624 from Microchip. This portion of the circuit is powered from a 110 VAC to 9 VDC wall cube. The linear regulator, LM7805, converts the 9 V output of the wall cube to 5 VDC. (MCP1525, Microchip; MCP3201, Microchip)

We have covered the analog section of this block diagram in most of the chapters of this book, all from different perspectives. This book has helped us learn the elements to good analog design so that this circuit is easy to implement. Figure 13.7 shows the final layout of this circuit.

Figure 13.7: This figure shows the layout for the circuit in Figure 13.6. Although this figure does not show the backside of this board, there is a complete, uninterrupted ground plane on that side.

Chapter 13

The layout implementation of this load-cell circuit (Figure 13.7) creates a noise-free, 12-bit result every time. Now the task is to include the digital portion of this circuit. Figure 13.8 diagrams circuit of the digital portion of the block diagram in Figure 13.5.

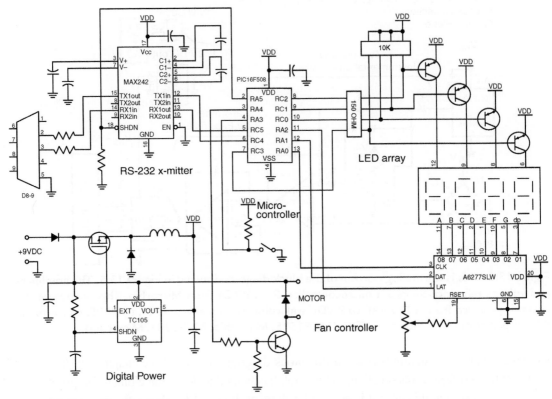

Figure 13.8: This is the circuit implementation of the digital portion of the block diagram in Figure 13.5. This circuit includes the microcontroller, RS-232 interface, LED array, fan controller and digital power.

The microcontroller is at the center of this circuit (Figure 13.8). Surrounding the controller, the RS-232 transmitter communicates directly with a laptop. You should exercise layout precautions because the voltages and consequently the digital noise is high with this portion of the circuit. The LED array generates high currents. EMI noise can be a potential if these traces are close to sensitive analog traces. The fan control circuit is also a potential EMI creator.

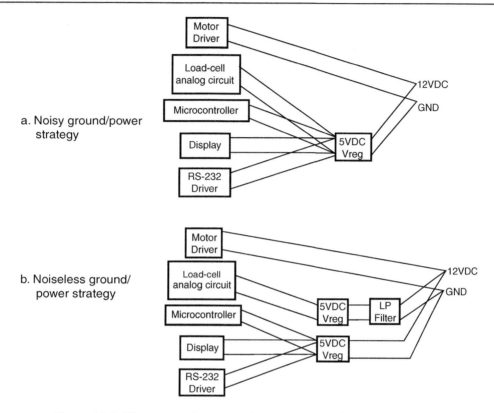

Figure 13.9: These are a few ground/power connection suggestions for the application circuit in Figure 13.5. The noise ground/power strategy (a) connects all 5 V systems to common power connections and the 12 volt noise driver (fan controller) to its own 12 V system. The conflict of the 5 V systems can cause circuit noise. Separation of analog and digital circuits will insure a lower noise environment.

When you consider adding digital circuitry to analog circuitry, first examine the power and ground strategies. Once you have done that, look at sensitive, analog, signal traces. Figure 13.9 illustrates a good way to examine your ground and power connections.

The optimum circuit separates the digital and analog power train and ground planes. Additionally, you need to separate the digital trace locations on the board with respect to analog traces. You will use the same analog circuit considerations when you add the digital circuit.

Chapter 13

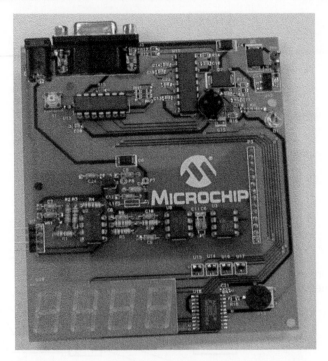

Figure 13.10: Combining analog and digital circuitry is easy if you keep track of your noise sources. The idea in this type of design is to keep noisy digital traces away from sensitive analog traces and devices.

Figure 13.10 shows a successful PCB implementation of the system shown in Figure 13.5. You can see that this board uses the analog layout shown in Figure 13.7. When under test, this board consistently produces 12-bit accuracy. This is due to the separation of analog and digital circuitry, a ground plane under that analog section, low noise analog devices, and by-pass capacitors.

Conclusion

In more recent years, I am not so opposed to new ideas as I once was. The digital engineers that I have worked with have helped me with this. They have a perspective that I have never experienced or imagined. That perspective was "digital" and all that goes with it. On a few occasions, when I started working with these engineers, I caught myself saying, "What! Are you crazy? That will never work." And in short order, these digital designers did make it work. So I have learned to listen, keep my mouth shut, and learn. It is working better, now.

APPENDIX A

Analog-to-Digital Converter Specification Definitions and Formulas

APPENDIX A

Analog-to-Digital Converter Specification Definitions and Formulas

Acquisition Time – The time required for the sampling mechanism to capture the input voltage, after the sample command is given for the hold capacitor to charge. Some converters have the capability of sampling the input signal in response to a sampling pin on the converter. With the current SAR CMOS converters they to sample with the clock after CS (chip select) drops (with SPI interfaces). Figure A.1 shows an example of this. In this case, it is nearly impossible to measure the acquisition time.

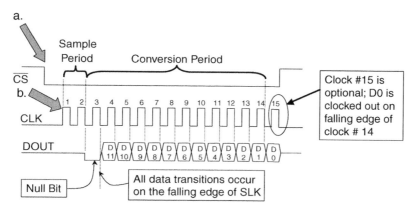

Figure A.1: (a) Chip select (CS) falls, the converter powers up and is becomes to sample. (b) The falling edge of the clock closes the sample switch and the converter starts to convert the signal.

Average Noise Floor – A calculated root-mean-square (rms) combination of the number of converter bits and the number of samples used in the calculation.

Bipolar Input Mode (Single-Ended or Differential Inputs) – An input range that uses two input pins and allows negative and positive analog inputs. A negative input is the difference between the two input pins. In this configuration, neither pin goes below or above the power supply rails.

Code Width – The voltage differential between two transition points. The code width is ideally equal to 1 LSB. See Figure A.2.

Appendix A

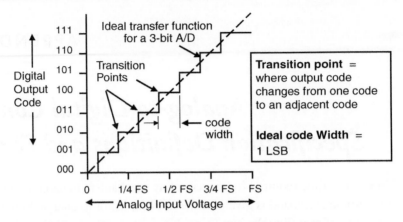

Figure A.2: The ideal transfer function has zero offset error, zero gain error, zero DNL error and zero INL error.

Common-Mode Rejection (CMR) – The degree of rejection of a common-mode signals (AC or DC) across the differential input stage.

Conversion Time – After sampling the signal, the time required for a SAR A/D converter to complete a single conversion.

Data Rate – The time required with a $\Delta\Sigma$ converter between the initiation of a conversion the availability of an output result.

Sigma-Delta Converter (Σ–Δ) – The sigma-delta converter is a one bit sampling system. In this system, multiple bits are sent through a digital filter where there is a fair degree of mathematical manipulation performed. With most industry converters, the digital filter is usually a finite impulse response (FIR) filter.

Differential Input – The A/D converter has two inputs per channel. These inputs subtract or differential the input signal, which provides a single digital output code per conversion.

Differential Nonlinearity (DNL) – The maximum deviation in code width from the ideal 1 LSB ($FS/2^n$) code width. You calculate the difference between all of the pairs of transitions in the transfer function. Figure A.3 illustrates the ideal transfer function as a solid line and the DNL error as a dashed line.

Digital Code Out = $2^n * V_{IN} / V_{FS}$

Where n: number of bits
V_{IN}: input analog voltage
V_{FS}: full-scale input voltage range

Digital Interface – SPI™ is a three or four-wire interface. With this interface, the chip is a slave device. I²C is a two-wire, Philips standard interface.

Analog-to-Digital Converter Specification Definitions and Formulas

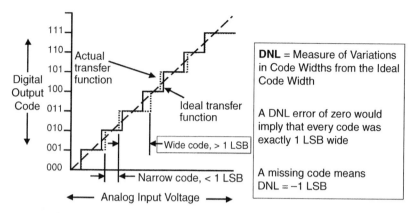

Figure A.3: The differential nonlinearity is the difference between an ideal code width and the measured code width.

Effective Number of Bits (ENOB) – The units of measure for SINAD is dB and the units of measure for ENOBs is bits. You can change SINAD into ENOB with the following calculation:

$$\text{ENOB} = (\text{SINAD} - 1.76)/6.02$$

Full-Scale Input (FS) – With A/D converters this input signal is analog. The full-scale input voltage is determined by the voltage reference value that is applied the converter reference pin. In many cases, the full-scale input range is equal to ground to the voltage reference value. In other cases, the full-scale input range is equal to ground to twice the voltage reference value. Refer to specific ADC data sheet for details.

Gain Error (Full-Scale Error) – The difference between the ideal slope between zero and full scale and the actual slope between the measured zero point and full scale. You zero out the Offset errors with this error calculation. See Figure A.4.

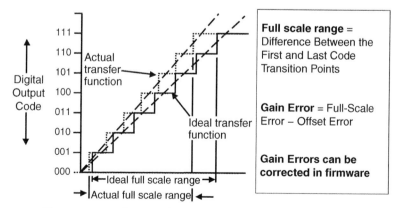

Figure A.4: Gain error is the difference between the ideal gain curve and the actual gain curve with offset removed.

Appendix A

Ideal A/D Converter Transfer Function – An analog voltage is mapped into an n-bit digital value with no offset, gain, or linearity errors. See Figure A.2.

Idle Tones – Caused by the interaction between the sigma-delta A/D converter modulator and digital filter. Idle tones usually occur with 2nd order modulators and 3rd order digital filters. They also occur around an analog input of zero volts plus or minus the converter's offset voltage. As the name implies, idle tones appear as a frequency in the output conversion with multiple DC input conversions at a constant data rate.

Integral Nonlinearity (INL) – The maximum deviation of a transition point from the corresponding point of the ideal transfer curve, with offset and gain errors zeroed. See Figure A.5.

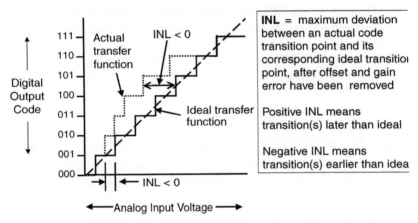

Figure A.5: INL is the aggregate of DNL errors and is equal to the maximum deviation for an ideal A/D converter transfer function.

Internal Buffer – The A/D converter input has a high impedance input that "isolates" the input signal from the converter.

Least Significant Bit (LSB) – The least significant bit is the bit representation of the smallest analog input signal that is converted. It is also referred to the furthest right bit in a binary digital word.

Monotonic – Implies that an increase (or decrease) in the analog voltage input will always produce no change or an increase (or decrease) in digital code. Monotinicity does not imply there are no missing codes. See Figure A.6.

Most Significant Bit (MSB) – The most significant bit is often thought as the furthest left bit in a binary digital word.

No Missing Code – Implies that an increase (or decrease) in the analog voltage input will always increase (or decrease) in digital output converter code. A converter with no missing code is also monotonic.

Analog-to-Digital Converter Specification Definitions and Formulas

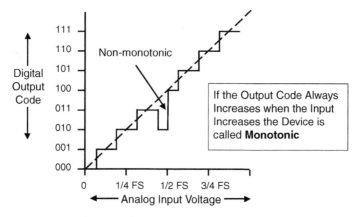

Figure A.6: This curve is nonmonotonic because an increase in the analog voltage can produce a smaller digital output code.

Normal-Mode Rejection – The attenuation of a specific frequency through the conversion process.

Number of Converter Bits (n) – The number of output codes of an A/D converter produces 2^n possible codes.

Offset Error – The difference between the first measured transition point and the first ideal, transition point. See Figure A.7.

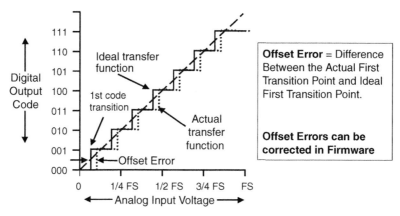

Figure A.7: Offset error is the difference between an Ideal 1st code transition and a measured 1st code transition.

Quantization Noise – The noise that an A/D converter generates as a consequence of dividing the input signal into discrete "buckets." The width of these "buckets" is equal to the LSB size of the converter. The quantization noise of a converter determines the maximum signal-to-noise ratio ($SNR_{IDEAL} = 6.02 \, n + 1.76$ dB).

Appendix A

Resolution – The number of possible output bits an A/D converter can produce in one conversion.

Sample and Hold – The analog switched input to a circuit with a sample and hold function open (samples) for a short duration to capture (hold) the analog input voltage.

Sampling Time – The time required to accurately sample an analog input signal.

Sample Rate – The speed that a converter can continuously convert several conversions. Typically specified as samples per second (sps) or Hertz (Hz)

Settling Time (as it relates to Sigma-Delta A/D Converters) – The settling time of the digital filter in a sigma-delta A/D converter reflects the order of the digital filter internal to the converter. This "time" is unit-less and equal to the number of filter stages.

Signal-to-Noise Ratio (SNR) – A calculated value that represents the ratio of signal power to noise power. The ideal SNR of an A/D converter is $6.02n + 1.76$ dB. For more information, refer to Appendix B (Reading FFTs).

Signal-to-Noise Ratio plus Distortion (SINAD or SNR+D) – The calculated combination of SNR and total harmonic distortion (THD). SINAD is the ratio of the RMS amplitude of the fundamental input frequency of the input signal to the RMS sum of all other spectral components below one half of the sampling frequency (excluding DC). The theoretical minimum for SINAD is equal to the SNR or $6.02n + 1.76$ dB. For more information, refer to Appendix B (Reading FFTs).

Single-Ended Inputs – An A/D converter that is configured for one input voltage that is referenced to ground.

Spurious Free Dynamic Range (SFDR) – The distance in dB on an FFT plot from the fundamental input signal to the first spur. For more information, refer to Appendix B (Reading FFTs).

Straight Binary Code – With the lowest input voltage, the digital count begins with all zeros and counts up sequentially all ones with a full-scale input. Straight binary is a digital coding scheme for unipolar voltages only.

Successive Approximation Register Converter (SAR) – The SAR converter uses a capacitive array at the analog input. You can manufacture this capacitive array and the remainder of the device in a CMOS process, making it easy to integrate it with microcontrollers or microprocessors.

Throughput Time – The time required for the converter to sample, acquire, digitize, and prepare for the next conversion.

Total Harmonic Distortion (THD) – The rms sum of the powers of the harmonic components (spurs) ratioed to the input signal power. For more information, refer to Appendix B (Reading FFTs).

Figure A.8: The straight binary code (also known as unipolar straight binary code) representation of zero volts is equal to a digital (0000). The analog full-scale minus one LSB digital representation is equal to (1111). With this code, there is no digital representation for analog full-scale.

Median analog voltage (V)	Digital code
0.9375 FS ($^{15}/_{16}$ FS)	1111
0.875 FS ($^{14}/_{16}$ FS)	1110
0.8125 FS ($^{13}/_{16}$ FS)	1101
0.75 FS ($^{12}/_{16}$ FS)	1100
0.6875 FS ($^{11}/_{16}$ FS)	1011
0.625 FS ($^{10}/_{16}$ FS)	1010
0.5625 FS ($^{9}/_{16}$ FS)	1001
0.5 FS ($^{8}/_{16}$ FS)	1000
0.4375 FS ($^{7}/_{16}$ FS)	0111
0.375 FS ($^{6}/_{16}$ FS)	0110
0.3125 FS ($^{5}/_{16}$ FS)	0101
0.25 FS ($^{4}/_{16}$ FS)	0100
0.75 FS ($^{3}/_{16}$ FS)	0011
0.1875 FS ($^{2}/_{16}$ FS)	0010
0.0625 FS ($^{1}/_{16}$ FS)	0001
0	0000

Total Unadjusted Error – The sum of offset, gain, and integral non-linearity errors.

Transition Point – The analog input voltage at which the digital output switches from one code to the next.

Two's Complement – See Figure A.9.

Figure A.9: The two's complement (also known as binary two's complement) representation of zero volts is also equal to a digital (0000). The analog positive full-scale minus one LSB digital representation is equal to (0111) and the analog negative full-scale representation is (1000).

Median Voltage (V)	Code
0.875 FS ($^{7}/_{8}$ FS)	0111
0.75 FS ($^{6}/_{8}$ FS)	0110
0.625 FS ($^{5}/_{8}$ FS)	0101
0.5 FS ($^{4}/_{8}$ FS)	0100
0.375 FS ($^{3}/_{8}$ FS)	0011
0.25 FS ($^{2}/_{8}$ FS)	0010
0.125 FS ($^{1}/_{8}$ FS)	0001
0	0000
−0.125 FS ($−^{1}/_{8}$ FS)	1111
−0.25 FS ($−^{2}/_{8}$ FS)	1110
−0.375 FS ($−^{3}/_{8}$ FS)	1101
−0.5 FS ($−^{4}/_{8}$ FS)	1100
−0.625 FS ($−^{5}/_{8}$ FS)	1011
−0.75 FS ($−^{6}/_{8}$ FS)	1010
−0.875 FS ($−^{7}/_{8}$ FS)	1001
−1 FS	1000

Unipolar Input Mode (Single-Ended Input) – An input range that only allows positive analog input signals.

Appendix A

Voltage Reference (also know as Analog Voltage Reference) – The input range (V_{IN}) and LSB sized is determined by the voltage reference (V_{REF}) to the converter. Depending on the converter, $V_{IN} = V_{REF}$ or $V_{IN} = 2V_{REF}$. LSB = $V_{REF} / 2^n$ or LSB = $2V_{REF} / 2^n$ (were "n" is the number of bits).

APPENDIX B

Reading FFTs

APPENDIX B

Reading FFTs

You would use the fast Fourier transform (FFT) tool to evaluate the ac performance of digitizing systems in the frequency domain. The theory of the Fourier series is somewhat complex, but the application is simple. The Fourier transform operates on the premise that you can reconstruct any signal or waveform by just adding together one or more pure sine waves with their appropriate amplitude, frequency, and phase.

For example, a square wave can be constructed from the Fourier series, sin(x) + 1/3 sin(3x) + 1/5 sin(5x) + 1/7 sin(7x). The addition of each element of this series, the fundamental pure sine wave (sin(x)), begins to transform into a square wave.

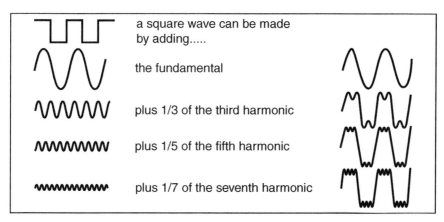

Figure B.1: You can construct a square wave using a fundamental sine wave and adding the odd harmonics of that sine wave.

Reading the FFT Plot

You generate an FFT plot by collecting a large number of digital samples from the output of the A/D, in a periodic fashion. Typically, A/D converter manufacturers use a single tone, full-scale analog signal, at the input of the A/D converter, for their typical performance curves for their specification sheets. Under these conditions, you exercise the full dynamic range of the converter. This data is then converted to the plot shown in Figure B.2. The frequency scale

Appendix B

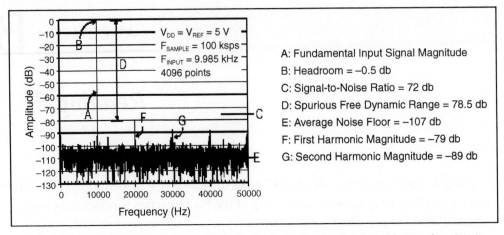

Figure B.2: Basic elements of the FFT plot include the fundamental input signal (A), signal headroom (B), signal-to-noise ratio (C), spurious-free dynamic range (D) and the average noise floor (E).

of this plot is always linear, from zero to nyquist/2. With FFT plots, the nyquist frequency is equal to the sampling frequency of the converter.

The magnitude axis ranges from zero down to an appropriate negative value, depending on the number of converter bits, and the number of samples included in the FFT calculation. When an analog input signal generates a full-scale output from the A/D converter, it will appear as zero dB on the FFT plot. Any magnitude less than full-scale can easily be converted into the digital code representation with these formulas:

$$D_{OUT} = (2^n - 1) * 10^{(MAGNITUDE / 20)}$$
$$V_{OUT\ RTI} = D_{OUT} * FSR / 2^n$$

> where, D_{OUT} is a decimal representation of the digital output code. D_{OUT} should be rounded to the nearest integer,
>
> MAGNITUDE is taken from the FFT plot and is in dB,
>
> $V_{OUT\ RTI}$ is a mathematical calculation that converts D_{OUT} into the same units as the analog input voltage. RTI = Referred to Input. This number should be equivalent to the analog input voltage, V_{IN},
>
> n is the number of A/D converter bits,
>
> FSR is the analog full-scale input range in volts

There are five elements of particular interest in the FFT plot, that provides insight into the system performance. Figure B.2 illustrates these five elements.

Fundamental Input Signal

The FFT plot in Figure B.2 uses the output signal of a 12-bit, SAR, A/D converter. The 12-bit converter has a sampling frequency of 75 kHz with a clock rate of 1.2 MHz. The analog input signal is 36 kHz (Figure B.2 (A)). A total of 4096 12-bit words are taken from the converter to generate this plot.

Input Signal Headroom

In reference to Figure B.2, the highest spur (A) represents the fundamental input signal to the converter. This signal exercises the converter's codes. In this case, the input signal is exercising the converter over as much of its input range as possible. The amplitude of the fundamental frequency in figure B.2 is 0.5 dB or 94.4% lower than full-scale, giving headroom (B) for the converter's output. This is done to insure that the converter is not overdriven, which will cause signal clipping. If signal clipping occurs, the FFT plot will show distortion of that signal in the form of spurs at frequencies other than the fundamental frequency.

Signal-to-Noise Ratio

A useful way of determining noise in the circuit of the A/D converter is with the signal-to-noise ratio (C). The signal-to-noise ratio (SNR) is a calculated value. It is the ratio of signal power to noise power. The theoretical limit of SNR is equal to $6.02n + 1.76$ dB, where n is the number of bits. An ideal 12-bit A/D converter should have a SNR of 74 dB. All spurs and the noise floor are included in the FFT calculation.

$$\begin{aligned}
\text{SNR} &= \text{rms Signal} / \text{rms Noise} \\
&= (\text{LSB } 2^{n-1}/ \sqrt{2})/(\text{LSB } \sqrt{12}) \\
&= 6.02n + 1.76 \text{ dB}
\end{aligned}$$

The SNR of the FFT calculation is a combination several noise sources. The possible noise sources include the quantization error of the A/D converter, internal noise of the A/D converter, noise from the voltage reference, differential non-linearity errors from the A/D converter and noise from the driving amplifier.

Spurious-Free Dynamic Range

The spurious-free dynamic range (D) quantifies the amount of distortion in the system. The spurious-free dynamic range (SFDR) is the distance from the fundamental input signal to the first spur (in dB).

Appendix B

Spurs resulting from the nonlinearity of the A/D converter will appear as a multiple (b) of the input signal's frequency (fundamental frequency), i.e., Asin(bx), unless they are a result of aliasing. If the spurs are a result of the aliasing phenomena, they are equal to:

$$f_{interference} = \pm (K f_{sample} - f_{aliased})$$

where $f_{interference}$ is the calculated possibilities of high frequency interference

K in a positive whole number
f_{sample} is the sampling frequency of the A/D converter
$f_{aliased}$ the aliased signal that appears on the FFT graph

In general, harmonically related spurs come from errors in the A/D converter. Non-harmonically related spurs are a result of other devices or external noise sources.

If the A/D converter creates spurs, it is probable that the converter has a degree of integral nonlinearity. The driving amplifier of the signal source can also create these spurs. The frequencies of these spurs are not related to the frequency of the fundamental frequency per the formula above. If the driving amplifier is the culprit it may have cross over distortion, be unable to drive the A/D converter, or be bandwidth limited. Injected noise can also cause these spurs from other places in the circuit, such as digital clock sources or the mains frequency.

Average Noise Floor

The average noise floor (E) in Figure B.2 is a combination of the number bits and the number of points used in the FFT. It is not a reflection of the performance of the A/D converter. Regardless the number of bits that the A/D converter has, the number of samples should be chosen so that the noise floor is below any spurs of interest.

$$\text{Average FFT Noise Floor (dB)} = 6.02\,n + 1.76\,\text{dB} + 10\log(3 * M / (\pi * ENBW)),$$

where M is the number of data points in the FFT.

ENBW is the equivalent noise bandwidth of the window function (see next section on how FFTs are generated)
n is the number of bits of the A/D converter

A reasonable number of samples for the FFT of a 12-bit converter is 4096.

Other Specifications from the FFT

There are two other specification of interest that the FFT calculations produce; total harmonic distortion (THD) and signal-to-noise plus distortion (SINAD). THD is the rms sum of the powers of the harmonic components (spurs) ratioed to the input signal power.

$$THD_{rms} = 20 \log (\sqrt{(10^{2nd\,HAR/20})^2 + (10^{3rdHAR/20})^2 + (10^{4thHAR/20})^2 + ...})$$

Significant integral, non-linearity errors of the A/D converter typically appear in the THD results. Most manufacturers specify THD by including the first nine harmonic components in this calculation.

SINAD is a calculated combination of SNR and THD where,

$$SINAD = -20\log(\sqrt{10^{-SNR/10} + 10^{+THD/10}})$$

FFT Accuracy

The FFT calculation is an effective tool to use in this situation. With an FFT, you can use the appropriate number of samples to calculate fairly reliable estimates. This sample size is associated with the level of "accuracy" or bits you are interested in accounting for. The formula below will give you good FFT results if the correct window is used:

FFT Accuracy (dB) = ± 4 dB / (n $\sqrt{K}$)   ; units in dB
FFT Accuracy (%) = $\pm 10^{(4/(20n\sqrt{K}))} - 1) \times 100\%$; units is %
Where n is the number of bits
    and K is the number of data point accumulated for the FFT

With this formula, you can determine how many samples you want to take as you evaluate your circuit noise. For instance, if I collect 256 samples from a circuit with a 12-bit A/D Converter I can only expect an FFT accuracy of 0.021dB or 0.24%. The accuracy of a good 12-bit converter is equal to $1/2^{12}$ or 0.024%; 10x better. I would say that 256 samples are not enough to make good noise decisions! A more suitable sample number would be 4096, which has an FFT Accuracy of 0.06%. Note that the square root of the number of samples, K, will prevent huge accuracy gains.

Windowing

Blackman/Harris – Bell-shaped window. Typically used for harmonic analysis of continuous time signals. Tapers data at ends of record to zero. Main lobe width is widest compared to other Windows. Has lowest adjacent side lobes as well as lowest farther side lobes from main lobe.

Hamming – Bell-shaped window. Typically used for harmonic analysis of continuous time signals. Tapers data to smaller values, but not zero, at ends of record. Has lower side lobes adjacent to main lobe than Hanning Window.

Hanning – Bell-shaped window. Typically used for harmonic analysis of continuous time signals. Tapers data at ends of record to zero. Side lobes farther from main lobe are lower than Hamming window.

Rectangular – Rectangular shaped window. Typically used for impulse response testing. Equivalent to multiplying all data record points by one. Gives best frequency resolution with narrowest lobe width. Amplitude accuracy errors occur if frequency of observed signal has a non-integer number of cycles in the FFT time record.

APPENDIX C

Op Amp Specification Definitions and Formulas

APPENDIX C

Op Amp Specification Definitions and Formulas

APPENDIX C

Op Amp Specification Definitions and Formulas

Absolute Maximum Ratings – Maximum conditions where the amplifier will fundamentally operate and not experience short term or long term damage. Amplifier specifications are not guaranteed under these conditions.

Beta (β) – The feedback factor in a closed-loop system. In simple systems, $1/\beta$ is the inverse of the closed-loop gain of the amplifier circuit from the noninverting input to the output. Beta is also a constant used to describe bipolar transistors. In this instance beta is the gain factor that describes the relationship of the collector, emitter, and base currents. In bipolar transistors $I_C = I_E + I_B$ and $I_C = (1 + \beta) * I_B$.

Closed-loop Output Resistance – Approximately equals the open-loop output resistance divided by the 1/beta or one over the feedback factor of the amplifier circuit.

Common-mode Input Voltage Range (V_{IN}) – See input voltage range.

Common-mode Rejection Ratio (CMRR) – The ratio of the differential change in the common-mode voltage to the resulting changes in offset voltage. $\text{CMRR} = 20 \log (\Delta V_{\text{COMMON-MODE VOLTAGE}} / \Delta V_{OS})$. CMRR is in decibels or dB.

Common-mode Input Resistance and Capacitance ($Z_{CM}, C_{CM-}, R_{CM-}, C_{CM+}, R_{CM+}$) – The effective resistance and capacitance between each input and ground. See Figure C.1.

Figure C.1: The common-mode resistance and capacitance on the input of the amplifier is in-between the input pins and ground.

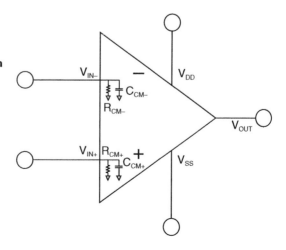

Appendix C

Differential Input Resistance and Capacitance (Z_{DIFF}, C_{DIFF}, R_{DIFF}) – The effective resistance and capacitance between the two inputs. See Figure C.2.

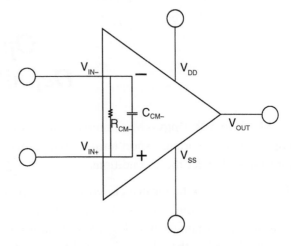

Figure C.2: Differential input impedance is the effective resistance and capacitance between the two input pins of the amplifier.

Full Power Response – The maximum frequency at which the amplifier can swing to the open-loop gain rated output voltage without significant distortion. The amplifier's slew rate limits this performance specification by inserting distortion as the signal frequency increases. See Figure C.3.

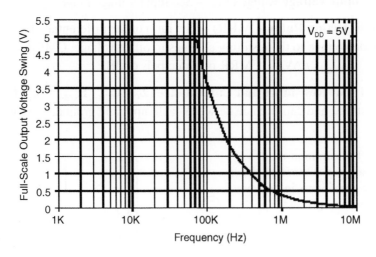

Figure C.3: The full power bandwidth of this amplifier is approximately 75 kHz.

Gain Bandwidth Product (GBWP) – The open-loop gain (V/V) times the frequency in the region where the open-loop gain is attenuating 20 dB/decade. For a unity-gain, stable amplifier the GBWP is equal to the 0 dB crossing of the open-loop gain curve.

Input Bias Current (I_{IN-} and I_{IN+}) – The current flowing in or out of the inverting (V_{IN-}) or noninverting (V_{IN+}) input terminals of an operational amplifier. This bias current originates in the input transistor, gate, or ESD cell structures.

Input Noise (Current and Voltage) – The noise that primarily originates in the input structure of the amplifier. Input noise is specified across a bandwidth. The units of input noise is Vrms, Vpeak-to-peak, amps rms, or amps peak-to-peak.

Input Noise Spectral Density (Current and Voltage) – A graphical representation of the culmination of spot noise across a frequency spectrum. The x-axis units are Hz; the y-axis units are V/√Hz or Amps/√Hz. See Figure C.4.

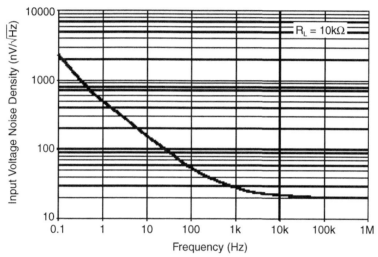

Figure C.4: A graphical representation of an amplifier voltage noise density over frequency.

Input Offset Current (I_{OS}) – The difference between the two input bias currents or ($I_{IN+} - I_{IN-}$).

Input Offset Current Drift (I_{OS}/TEMP) – The rate of change of the input bias current with temperature.

Input Offset Voltage (V_{OS}) – The differential DC input voltage required to provide a zero output voltage.

Input Offset Voltage Drift (V_{OS}/TEMP) – The rate of change of the input offset voltage with temperature.

Input Voltage Range (V_{IN}) – The maximum and minimum input voltages at which the amplifier will operate in its linear region as defined by the open-loop gain specification.

Operating Temperature Range – Temperature range where the amplifier will operate but not necessarily meet specifications called out in the table of specifications.

Appendix C

Open-loop Gain (A_{OL}) – The magnitude of the gain of the operational amplifier block. Open-loop Gain lessens with increasing frequency. $A_{OL} = 20 \log (\Delta V_{OUT} / \Delta V_{OS})$. The units of measure for Open-loop gain is dB or volts/volts.

Output Resistance – See Closed-loop Output Resistance and Open-loop Output Resistance.

Output Swing Bandwidth – See Full Power Response.

Output Voltage Swing (V_{OUT}) – The maximum and minimum output voltages that under specified load conditions.

Phase Margin – The phase of the open-loop gain at its 0dB crossing plus 180 degrees, with respect to the noninverting input of the amplifier. Theoretically, a phase margin greater than zero degrees indicates a stable system. In practice, a phase margin must be greater than or equal to 45°.

Power Supply Current (I_{DD}) – See Quiescent Current.

Power Supply Rejection Ratio (PSRR) – The ratio the difference of two power-supply setting to the change in offset voltage. $PSRR = 20 \log (\Delta V_{POWER\ SUPPLY} / \Delta V_{OS})$. PSRR reduces with increasing frequency.

Power Supply Voltage – Allowable voltage range for the positive power supply in reference to the negative power supply.

Quiescent Current (I_{DD}) – Amount of current that will be sourced from the positive power supply or sinked into the negative power supply with no load on the output of an amplifier.

Rated Output – See Output Voltage Swing.

Settling Time – Given a step input, the time required for the output voltage to settle within a specified percentage of its full value.

Storage Temperature Range – Maximum temperature range for storage of the operational amplifier. If the temperature exceeds this range, damage to the amplifier may occur.

Specified Temperature Range – Temperature Range where the amplifier will meet specifications called out in the table of specifications.

Short-Circuit Current (I_{SC}) – When the output is shorted mid-between the power supply this is the maximum output current.

Slew Rate (SR) – The maximum rate of change of the output voltage.

Spot Noise – Noise amplitude within a bandwidth of 1 Hz.

Unity Gain Bandwidth – The frequency range from DC to the frequency at which the open-loop gain crosses unity.

Index

Symbols
.OPTION (.OP) statement, SPICE, 175–78
1/f noise, 241, 243–44, 247

A
absolute maximum ratings, 339
acquisition time, 321
active filter, 95, 97–8, 105–6, 109, 142–3, 147–8, 312
AC induction motor, 65, 68, 83–5, 87
ADC
  accuracy, 33–5, 38–9, 42, 44–5, 50, 56, 200
  effective number of bits (ENOB), 35–7
  effective resolution (ER), 35–7
  error analysis, 203
  full-scale range (FSR), 28
  gain error, 31, 35, 38–9, 46, 56
  input range, 28–9, 38
  offset error, 31, 35, 38–9, 46, 56
  repeatability, 35, 38
  resolution, 33–5, 38, 47, 52–3, 56, 199–200, 202
  settling time, 57, 58
  signal to noise ratio (SNR), 34–6, 52–3
  signal to noise ratio plus distortion (SINAD), 35–6
  throughput rate, 33–4
Alexander, Bowers macromodel, 177
alkaline, 212
amplifier
  design pitfalls, 131
  input capacitance, 152
  input stage, 138–41, 144, 148, 150–1
  output distortion, 144–6
  output stage, 146, 148, 152

analog filter, 4, 8, 10, 14, 20, 70–1, 73–81, 83, 88, 93–5, 98, 104–5, 108, 110, 312
analog ground (AGND), 266, 275–7, 280, 286
anti-alias, 95, 103, 105, 109–15
anti-alias filter, 43–4, 49, 70–1, 73–81, 83, 88, 95, 142–3, 147–8, 238, 258–9, 286–8, 310
approximation types, 95, 98
auto router, 277–8
average noise floor, 321, 332, 334

B
battery chemistry, 211–2
Bessel, 95–6, 98, 100–3, 110, 112–3
beta (β) 1/β, 155–61, 339
binary two's complement, 29, 31–2
bipolar amplifiers, 121
bipolar input mode, 321
bipolar transistors, 84
Blackman/Harris window, 335
board
  capacitance, 267–8
  inductance, 265, 267, 274
  resistance, 267, 289
Bode plot, 152–3, 156–7, 161
Boltzman's constant, 239
Boyle, Pederson macromodel, 181
breadboard, 294, 304
brick wall, 96
bridge sensor, 11
broadband noise, 243
Buck-SPC, 215–6, 219–20
buffer amplifier, 121–4, 140–2, 148, 150, 162, 313
Butterworth filter, 95–6, 98–101, 110, 112
bypass capacitor, 237–8, 255–9, 263–5, 276, 286–7, 289

Index

C

CANbus, 4–5
charge pump, 213–6, 219–24, 228, 239, 250, 252
Chebyshev filter, 95–6, 98, 100–1, 110–2
clock start-up time, 225–6
closed-loop amplifier system, 154–5
closed-loop noise gain, 246
closed-loop output resistance, 339
CMOS amplifiers, 121, 123, 129, 132
code width, 321
common-mode input resistance and capacitance (Z_{CM}, C_{CM-}, R_{CM-}, C_{CM+}, R_{CM+}), 245, 339
common-mode input voltage range (V_{IN}), 339
common-mode rejection (CMR), 322
common-mode rejection ratio (CMRR), 148–51, 339
comparator, 187, 194–7, 199–202, 204–7
components location, 263, 267, 284
conducted noise, 233–4, 238–9, 252, 254, 259
constant current source, 72–4, 76, 172–3
conversion time, 322
crest factor, 234–5, 239, 244
current-to-voltage conversion, 122, 128, 310, 313
current return path, 277, 280–2, 285, 288
cut-off frequency, 94–7, 100, 108–111

D

DAC
  accuracy, 189, 192, 194
  resolution, 189, 192, 194
data rate, 322
DC operating point, 175–6
dependent sources, 179, 182
device noise, 233–4, 238–9, 249, 252–4
difference amplifier, 122, 126–7, 129–30, 313
differential input, 28–9, 33, 40, 50, 321–2
differential input resistance and capacitance (Z_{DIFF}, C_{DIFF}, R_{DIFF}), 245, 340
differential nonlinearity (DNL), 31, 35, 38–9, 322
digital-to-analog converter (DAC), 188–90, 193, 200–1, 269–70, 272–3
digital code out, 322
digital filter (decimation, FIR, IIR), 11, 13, 47–8, 50–2, 54, 56–8, 93–5, 312
digital ground (DGND), 275–8, 280, 286
digital interface, 322

digital potentiometer, 269–73
digital signal processor (DSP), 84

E

effective number of bits (ENOB), 35–7, 323
effective resolution (ER), 35–7
efficiency, 213–7, 219–23
electro-magnetic interference (EMI), 213, 221, 265–6, 274, 277, 281, 286, 316
energy density, 212–3

F

Fast Fourier Transform (FFT), 190–1, 293–4, 298–303, 305, 314, 335
ferrite bead, 237, 252, 258–9
filter pass band, 96–7, 99–102, 113–4
filter stop band, 96, 98, 101–2, 111
filter transition band, 96, 98, 100–2, 104
FIR filter, 95
floating current source, 131, 173
full-scale error, 323
full-scale input (FS), 323
full power response, 340
fundamental input signal, 332–3

G

gain bandwidth product (GBWP), 8–11, 105, 108–9, 340
gain error, 31, 35, 38–9, 46, 56, 323
GMIN, 178
ground plane, 266–7, 275–7, 280–2, 284–6, 288, 314, 318
ground trace, 265–6, 269, 278, 281

H

Hall effect sensor, 83, 86, 88
hamming window, 335
hanning window, 335
headroom, 332–3
high-pass filter, 93, 105
histogram, 295–6, 301, 314
hysteresis, 194–6

I

I/O gates or I/O ports, 187–8, 190, 199, 207
ideal A/D converter transfer function, 324

ideal op amp, 120
idle tones, 324
IGBT, 84
IIR filter, 95
independent sources, 179
input-offset distortion, 137–8, 141–2, 148
input bias current (I_{IN-} and I_{IN+}), 341
input capacitance, amplifier, 152
input noise (current and voltage), 341
input noise spectral density (current and voltage), 341
input offset current (I_{OS}), 341
input offset current drift (I_{OS}/TEMP), 341
input offset voltage (V_{OS}), 341
input offset voltage drift (V_{OS}/TEMP), 341
input stage, amplifier, 138–41, 144, 148, 150–1
input voltage noise, 237–8, 243–4
input voltage range (V_{IN}), 341
instrumentation amplifier, 11–3, 69–70, 72, 79–80, 129–30, 140, 142, 236, 238, 241–3, 253–4, 259, 310
integral nonlinearity (INL), 31, 35, 38–9, 324
integrated silicon temperature sensor, 66–7
internal buffer, 324
inverting gain amplifier, 122, 125

L

least significant bit (LSB), 324
LED display, 314, 316
Li-Poly, 212
Lithion-Ion, Li-Ion, 211–3, 221, 223
lithium, 212
load-cell circuit, 283, 297, 301, 314, 316
low-pass filter, 70–1, 73–81, 83, 88, 93, 95–97, 99–101, 103–8, 110–15, 142–43, 147–48, 189–94, 200, 238, 258, 259, 312
low dropout regulator (LDO), 213–4, 217–22, 250–1, 253–4

M

macromodel, 167, 169–70, 175–83
maximally flat, 98–9, 101
mean, 234–5
microcontroller Σ–Δ converter, 194, 199–201, 203, 205–7
metal oxide semiconductor field effect transistor (MOSFET), 68, 83–6, 216, 218
monotonic, 324
most significant bit (MSB), 324
motor control, 65, 68, 83–5, 87
multimeter, 293–5, 314
multiplexer, 95, 99, 112–3, 162–3
multiple feedback low-pass filter, 108–9

N

negative temperature coefficient (NTC) thermistor, 6–7
NiCd, 212–3
NiMH, 212–3
noise, 28–9, 31, 33, 35, 312–4, 317–8
noise-shaping filter, 50, 52–3
noise statistics, 36–7
noninverting gain amplifier, 122, 125–6
normal-mode rejection, 325
normal distribution, 234–5, 237
no missing code, 324
number of converter bits (n), 325
Nyquist, 94, 103

O

offset error, 31, 35, 38–9, 46, 56, 325
offset voltage, 138, 140–1, 144, 148–51
open-loop gain (A_{OL}), 148, 150, 152–8, 162, 246–7, 342
operating temperature range, 341
operational amplifier noise, 243, 247–9, 253–4
optocoupler, 83, 87–8
op amp input stage, 119, 121, 129
op amp stability, 16–9, 173, 175
oscilloscope, 293–4, 302, 314
output distortion, 144–6
output resistance, 342
output stage, 146, 148, 152
output swing bandwidth, 342
output voltage swing (V_{OUT}), 342
overshoot, 97–9, 101, 103, 112, 114–5, 159, 162–3

P

parasitic capacitance, 168–9, 174–5
passive filter, 95, 105–6
peak-to-peak noise, 235, 252

Index

perf board, 304
phase margin, 97, 152, 154, 157–63, 342
phase shift, 152, 154, 157–63
photodetector, sensing, diode, 65, 67, 81–3, 114
photodiode pre-amp, 146–7
poles, 95–101, 106–8, 112
power connection plan, 317
power meter, 8
power plane, 263, 266, 27–7, 284–6, 314
power supply current (I_{DD}), 342
power supply filter, 254, 257–8
power supply noise, 250, 252, 256–8
power supply rejection ratio (PSR) (PSRR), 148–51, 342
power supply trace, 26–6, 284
power supply voltage, 342
pressure sensor, 69, 78–80, 295–7, 301
primary battery cell, 212
propagation delay, 97, 115
pseudo-differential input, 28
pulse frequency modulation (PFM), 252
pulse width modulation (PWM), 252
pulse width modulator (PWM), 84–6, 187–94
PWM duty cycle, 188

Q

quantization noise, 325, 333
quiescent current, 342

R

radiated noise, 233–4, 254
radio-frequency interference (RFI), 277
random noise, 234, 239, 298–9, 301
rated capacity, 212
rated output, 342
rectangular window, 335
referred-to-input (RTI), 243, 247–8, 251
referred-to-output (RTO), 248
regulated charge pump, 219, 228
  repeatability, ADC, 35, 38
resistance temperature detector (RTD) sensor, 66–8, 72–7, 95, 109–10
resistor noise, 239–42, 253–4
resolution, 326
  resolution, ADC, 33–5, 38, 47, 52–3, 56, 199–200, 202

resolution, DAC, 189, 192, 194
ripple, 96, 100–1, 112
root-mean-square (rms), 234–5, 239–40, 244, 313
RS-232 transmitter, 316

S

Sallen-key low-pass filter, 107–9
sample and hold, 326
sample rate, 326
sampling time, 326
SAR converter layout, 274–6
SAR converter, 311
secondary cell, 212–3
settling time, 57–8, 162, 326, 342
short-circuit current (I_{SC}), 342
sigma-delta ADC (Σ–Δ), 194, 199–201, 203, 205–7, 311, 313, 322
sigma-delta board layout, 274, 276–7
signal-to-noise ratio (SNR), 34–6, 52–3, 249–50, 326, 332–3
signal-to-noise ratio plus distortion (SINAD or SNR+D), 35–6, 326, 334–5
single-ended input, 28–9, 321, 326–7
slew rate (SR), 8, 10–1, 145, 162, 342
specified temperature range, 342
SPICE simulation, 241–2, 248–9, 313
spot noise, 342
spurious free dynamic range (SFDR), 326, 332–3
stability analysis, 152, 157
standard deviation, 234–5
star layout configuration, 281–2
storage temperature range, 342
straight binary, 28–30, 32
straight binary code, 326–7
successive approximation register (SAR) converter, 311, 326, 333
summing amplifier, 122, 127–8, 313
switched-capacitor filter, 70–1
switched power converter (SPC), 213–6, 219–22, 251–2

T

temperature sensors, 66–7, 69, 72–3, 76–7
thermistor, 66–7, 72
thermocouple, 66–8
throughput rate, ADC, 33–4

throughput time, 326
timer, 187, 194, 197, 199–200
time domain, 300, 303
total harmonic distortion (THD), 326, 334
total unadjusted error (TUE), 327
touch screen, 65, 68–9
transimpedance amplifier, 17, 121, 146–7
transistor level model, 170–1, 179–82
transition point, 327
two's complement, 327
two-clock start-up, 222, 227
two-layer board, 277, 289

U

unipolar input mode, 327

unipolar straight binary, 28–30, 32
unity gain bandwidth, 342

V

voltage follower amplifier, 122, 140–2, 148, 150, 162
voltage reference, 328

W

wall cube, 238, 252, 257, 315
Wheatstone bridge, 69, 78
window comparator, 187, 196–7